Biofertilisers : Commercial Production Technology and Quality Control

Biofertilisers : Commercial Production Technology and Quality Control

Editor

Abhijeet Sawant

scitus academics

Biofertilisers : Commercial Production Technology and Quality Control

Edited by **Abhijeet Sawant**

ISBN: 978-1-68117-049-7

Library of Congress Control Number: 2015935444

SCITUS Academics LLC,
www.scitusacademics.com
Box No. 4766, 616 Corporate Way,
Suite 2, Valley Cottage,
NY 10989

Printed in United States of America on Acid Free Paper ∞

Contents

Preface

Man made fertilizers have come to stay in the 20 century for increasing the output of agricultural production and to meet the ever increasing demand of human population, which has been further accentuated by the limited availability of additional fertile farm land. The pathway of productivity improvement followed so far has, however, been accompanied by an exponential increase in the consumption of nonrenewable form of energy. Manufacture of chemical fertilizers is heavily dependent on energy derived from fossil, which is getting depleted at a very fast rate. On the contrary, biofertilizers require much less energy and are dependent on energy from renewable sources such as products of photosynthesis and organic matter in the soil. The nutrients provided by biofertilizers to the agroecosystem do not pose any health risk to human which is being seriously apprehended by the use of chemical synthetic fertilizers. With almost twice the quantity of plant nutrients being removed from soil than added through fertilizers, the growing nutrient imbalance poses a major threat to sustain soil health and crop productivity. Huge investment will be needed to install factories to meet the growing demand of chemical fertilizers. The fertilizers so produced is not only high price but also the factories themselves have and adverse effect on the environment. Chemical fertilizers are not only costly but growing awareness of environment pollution and limitations of nonrenewable resource may introduce additional constraints. Since the ancient time, mankind has taken advantage of microorganism appreciating their beneficial role in living system. In the later half of last century, their role in agriculture has been recognized as biofertilizers. The study on microbes due to commercialization of biofertilizers in world market has taken a new turn for their high tech mass production. Biofertilizer inoculants are easy for augmentation due to simplicity in production technology developed in recent years. They are natural allies and effective for integrated nutrient management systems with a high degree of safety not only to human and plant lives but also for environmental compatibility. Incorporation of biofertilizers in combination with organic manures and inorganic fertilizers can completely obviate the detrimental effects of current agricultural practices.

Editor

Chapter 1

Technologies for Beneficial Microorganisms Inocula used as Biofertilizers

E. Malusá[1, 2], L. Sas-Paszt[1], and J. Ciesielska[1]

[1]Research Institute of Horticulture, 96-100 Skierniewice, Poland
[2]CRA, Centre for Plant-Soil Systems, Turin, Italy

ABSTRACT

The increasing need for environmentaly friendly agricultural practices is driving the use of fertilizers based on beneficial microorganisms. The latter belong to a wide array of genera, classes, and phyla, ranging from bacteria to yeasts and fungi, which can support plant nutrition with different mechanisms. Moreover, studies on the interactions between plant, soil, and the different microorganisms are shedding light on their interrelationships thus providing new possible ways to exploit them for agricultural purposes. However, even though the inoculation of plants with these microorganisms is a well-known practice, the formulation of inocula with a reliable and consistent effect under field conditions is still a bottleneck for their wider use. The choice of the technology for inocula production and of the carrier for the formulation is key to their successful application. This paper focuses on how inoculation issues can be approached to improve the performance of beneficial microorganisms used as a tool for enhancing plant growth and yield.

INTRODUCTION

Environmental issues such as freshwater pollution, energy saving, and soil erosion are forcing the farmers to introduce methods of cultivation that have a lower impact on the environment. The application of environmentaly friendly practices is promoted by voluntary certification schemes (e.g., GlobalGAP or organic farming schemes) as well as by legally binding regulations (e.g., the EU Directive 2009/128 aiming at the implementation of sustainable pest management practices). In this context, the reduced use of chemical fertilizers with increased application of organic fertilizers is considered a compulsory route to alleviate the pressure on the environment derived from agricultural practices.

Several organic fertilizers have been introduced in the recent years, which are also acting as natural stimulators of plant growth and development [1, 2]. A specific group of this kind of fertilizers includes products based on plant growth-promoting microorganisms (PGPM). Three major groups of microorganisms are considered beneficial to plant nutrition: arbuscular mycorrhizal fungi (AMF) [3], plant growth-promoting rhizobacteria (PGPR) [4], and nitrogen-fixing rhizobia, which are usually not regarded as PGPR [5]. Microbial inoculants based on these microorganisms can be divided into different categories depending on their use, even though exact definition of these categories is still unclear. Nevertheless, the category of biofertilizer most commonly refers to products containing soil microorganisms increasing the availability and uptake of mineral nutrients for plants (like rhizobia and mycorrhizal fungi). According to the definition proposed by Vessey [6], biofertilizers are substances which contain living microorganisms which, when applied to seed, plant surfaces, or soil, colonize the rhizosphere or the interior of the plant, and promote growth by increasing the supply or availability of primary nutrients to the host plant. Another category of PGPM-containing products is that of phytostimulators which are generally containing auxin-producing bacteria, inducing root elongation [7].

The interest in the application of these products is rising due to the enhancement in nutrient uptake efficiency [8, 9] and society demands for more green technologies in production [10], increasing costs of agrochemicals. Furthermore, biofertilizers and phytostimulators possess secondary beneficial effects that would increase their usefulness as bioinoculants. Indeed microorganisms such as Rhizobium and Glomus spp. have been shown to also play a role in reducing plant diseases [11].

The practice of inoculating plants with PGPM can be traced back to the early 20th century, when a product containing Rhizobium sp. was patented (Nobbe and Hiltner 1896, cited in [12]). Mycorrhizal fungi, even though utilized as biofertilizers since few decades, were reported to promote plant growth through P uptake since the late 1950s [13]. Since then, research efforts in these fields have steadily increased, resulting, in recent years, in the selection of numerous strains showing several beneficial features [4, 6, 13–16].

The policies supporting sustainable agricultural production and extensive research that has improved the effectiveness and consistency of microbial inocula have resulted in the registration of several strains for both biocontrol [17] and biofertilization [4], with mycorrhizal and PGPR preparations being marketed in several countries. Yet, a wider use of microbial inoculants, especially those acting as phytostimulators and biofertilizers, has been frequently hampered due to the variability and inconsistency of results between laboratory, greenhouse, and field studies. The reason for these discrepancies lies in the incomplete understanding of the complex relationships established between the components of the system: the plant, the microorganisms, and the environmental conditions, particularly that of soil [18]. In addition, the lack of correct formulations and the expensive and time-consuming procedures of registration are also among the factors holding back the use of PGPM on a wider scale [19].

The present paper is focusing at different issues related to formulation of inoculants to improve the performance of PGPM use in agriculture, particularly for the purpose of plant nutrition.

INOCULATION TECHNOLOGY

PGPM inoculants can be defined as formulations containing one or more beneficial microorganism strains (or species) prepared with an easy-to-use and economical carrier material. The development of techniques for the production of large quantities of pure inocula, with high infectivity potential, is the main issue to be tackled in order to allow a wide use of biofertilizers. The key aspects in PGPM inoculation technology are the use of a proper formulation of inocula preparations, the selection of an adequate carrier, and the design of correct delivery methods.

Inocula Formulation

The production of selected bacteria and yeasts in pure cultures is a quite common practice making use of fermenters. Therefore, once the particular strain/s for the inoculum have been selected, an industrial standardized process of production can be defined [20]. However, in case of biofertilizers, unlike that of biopesticides, the cost of production is an important constraint considering that the price of the fertilizer shall not exceed that of conventional ones to assure a market sustainability. Hence, several cheap organic matrixes (e.g., whey, water sludges, composts, etc.) have been tested as growth media for PGPM [21, 22]. Another approach to reduce the production costs is by using agroindustrial residues enriched with rock phosphate. During composting or fermentation, free or immobilized microorganisms that produce organic acids are added to the matrix, improving the solubilization of phosphate, which make it more available to plants [23].

Recently, the use of biofilms has also been proposed as possible means to produce effective plant inocula [24]. A biofilm consists of microbial cells embedded into a self-produced polymeric matrix (known as an extracellular polymeric substance—EPS) and adherent to an inert or living surface, which provides structure and protection to the microbial community. Three major types of biofilms can occur in the soil: bacterial (including Actinomycetes), fungal, and fungal-bacterial biofilms. Both bacterial and fungal biofilms are formed on abiotic surfaces, while fungi act as the biotic surface in formation of fungal-bacterial biofilms [24]. The majority of plant-associated bacteria found on roots and in soil are forming biofilms [25]. Therefore, using PGPM strains that are forming biofilms could be a strategy to ease the formulation and production of inocula. Furthermore, biofilm-based inocula could also facilitate the production of biofertilizers considering the biofilm as a carrier (see below).

While ectomycorrhizal fungi can be produced under fermentation conditions, the production of AMF inocula poses several difficulties due to the need of a plant host for the multiplication of the mycorrhizal fungi. The first attempts in AMF inocula production used pot cultures with soil mixtures, or other technologies (such as aeroponics) [13, 26]. However, the development of monoxenic cultures in the late 1980s [27, 28] has allowed the production of AMF under strictly controlled conditions. A method utilizing split-plate cultures and Ri T-DNA transformed roots of carrots [29] was developed to produce spores. Even though the method allows a higher efficiency with production on average of 15.000 spores per Petri dish 4-5 months after beginning the production cycle, it has been used mainly for physiological and laboratory studies. The improvement of this method proposed by Douds [30] requires replacing the media in the distal compartment every 2 months and concomitantly replenishing the carbon source in the proximal compartment with glucose. The results are a production of about 65.000 spores in 7 months. Yet, such methods are mainly used for the production of batches of spores for trials or for maintenance of genebanks since the annual cost for producing one spore was estimated to be up to 30–50 USD, depending on the method utilized [31]. Recently, a large-scale in vitro production of mycorrhizal fungi, feasible for implementation on a commercial scale, has been proposed [32]. It emphasizes the selection of appropriate Ri T-DNA transformed host roots for different AMF species, the choice and maintenance of the growth medium, and the application of quality assurance procedures.

However, commercial inoculants containing AMF species are still produced mainly by growing host plants in controlled conditions, with the inclusion in the inoculant of different fungal structures (spores, mycelium hyphae) and containing mycorrhizal roots residues from the plants used as the propagating material (i.e., sorghum, maize, onion, or Plantago lanceolata). This could be considered a classical method where substrates of sand/soil and/or other materials (e.g., zeolite, perlite) are used to mass-produce AM fungal inoculum in pots, bags, or beds, for large-scale applications. Critical aspects of this production method are [33]

- The use of known AMF species [34],
- The choice of the host species with a short life cycle, adequate development of the root system, a good colonization level by a large range of AM fungi, and tolerance to relatively low levels of phosphorus,
- The manipulation of mineral nutrients level in soil,
- The appropriate combination of AMF species and host plant.

With this technique, it is possible to reach inoculum densities of 80–100 thousand propagules per litre [35]. This implies the need of diluting the inoculum with a carrier for the preparation of a commercial product (see below).

Considering that microbial associations between bacteria and mycorrhizal fungi have been observed to occur naturally in the soil promoting mycorrhizal symbiosis [36, 37], enhanced formulations could include two or more species of different PGPM. Microbial consortia can stimulate plant growth through a range of mechanisms that improve nutrient acquisition and inhibit fungal plant pathogens [18, 38]. The different mechanisms proposed to explain such growth stimulation relate to the increased rate of nutrients cycling due to enhanced soil microbial content and microorganisms biodiversity found in soil where mycorrhizal plants are grown [37, 39].

Simultaneous inoculation with different PGPR and/or AMF often resulted in increased growth and yield, compared to single inoculation through improved nutrient uptake [40, 41]. Indeed, the interactions between bacteria and AM fungi have beneficial functions related to nutrient uptake, particularly when PGPR [42–44] and N_2-fixing bacteria [45, 46] are involved.

Inoculation of maize and ryegrass with A. brasilense and AMF resulted in N and P contents comparable to plants grown with fertilizer [8]. Coinoculation with different AMF species is generally more likely to be effective due to the general not specificity of AMF fungi colonization of specific plant species/cultivars [47, 48]. Synergistic interaction between AM fungi and several PGPR, including Azospirillum, Azotobacter, Bacillus, and Pseudomonasspecies, has also been reported as beneficial for plant growth (see also the review by Barea et al. [49]). Increased root colonization by AMF was observed when mycorrhizal fungi were coinoculated with PGPR [50, 51]. Four times higher nodule number was observed when plants were inoculated with a mixture containing Glomus deserticola and Rhizobium trifoli, in comparison to single R. trifoli, inoculation, and enhanced mycorrhization and nodulation was obtained with coencapsulated R. trifoli and Yarrowia lipolytica [52]. Inoculation with nodule-inducing rhizobia and AM fungi resulted in increasing both P and N uptake efficiency [53]. Mycorrhizal and nodule symbioses often act synergistically on infection rate, mineral nutrition, and plant growth [54]. Coinoculation resulted in enhanced uptake of mineral nutrients and increased growth [55, 56] also when PGPM were applied as commercial biofertilizers containing consortia of different microorganisms [9, 57–60].

All these examples are pointing to the usefulness and higher efficacy of biofertilizers composed by more species having different mechanisms of growth promotion. The availability of several strains of PGPR [15] and AMF [33] tested in different crops species and under different field conditions should allow the definition of consortia suitable for commercial uses.

Carriers

The carrier is the major portion (by volume or weight) of the inoculant that helps to deliver a suitable amount of PGPM in good physiological condition [61]. The materials constituting the carrier can be of various origins: organic, inorganic, or synthesized from specific molecules. Availability and cost are the main factors affecting the choice of a carrier.

The carrier should be designed to provide a suitable microenvironment for the PGPM and should assure a sufficient shelf life of the product (at least 2-3 months for commercial purposes, possibly at room temperature). The formulation should allow an easy dispersion or dissolution in the volume of soil near the root system. A good carrier should therefore possess as much as the following properties: good moisture absorption capacity, easy to process and free of lump-forming materials, near-sterile or easy to sterilize by autoclaving or by other methods (e.g., gamma-irradiation), low cost and availability in adequate amounts, and good pH buffering capacity

[62]. For carriers that shall be used for seed coating, a good adhesion to seeds is also important [63]. Other characteristics that are affecting the carrier appropriateness are a standardized composition ensuring chemical and physical stability, suitability for as many PGPM species and strains as possible, the possibility of mixing with other compounds (i.e., nutrients or adjuvants), and being composed of biodegradable and nonpolluting compounds [61]. In case the inoculant is used as seed coating, the carrier shall assure the survival of the PGPM on the seed since normally seeds are not immediately sown after seed coating [64]. Survival of the PGPM is important both during the storage period of the bioproduct and after being introduced into the soil [65]. The latter is fundamental for defining the application technology and dosing the product: the inoculant has to compete with native soil microorganisms for the nutrients and habitable niches, and has to survive against grazing protozoa [66, 67]. Carrier materials that make available nutrients and/or habitable micropore to the PGPM, particularly in case of bacteria, would then be more suitable.

The kind of carrier utilized defines the physical form of the biofertilizer. Dry inoculants can be produced using different kinds of soil materials (peat, coal, clays, inorganic soil), organic materials (composts, soybean meal, wheat bran, sawdust, etc.), or inert materials (e.g., vermiculite, perlite, kaolin, bentonite, silicates) [61]. Liquid inoculants can be based on broth cultures, mineral or organic oils, or on oil-in-water suspensions.

In the case of solid carriers, powder, granules, or beads are the most typical forms utilized. Standard sizes of the powder material may vary from 75 μm to 0.25 mm [61]. The size of granules and beads ranges from 100–200 μm to 3-4 mm in diameter [12]. Powder-type inoculants can be used to coat seeds or suspended in a liquid to form a slurry that is directly applied to the furrow or, alternatively, the seeds/plants are dipped in it just prior to sowing/planting [68].

Bacteria can also be stored by lyophilization, which allows achieving high survival rates [69], without any carrier. However, during the process a cryoprotectant must be added, which is essential for protecting the bacterial cell membrane and cytoplasm against dehydration. Mannitol is a good protectant, but recently microcrystalline cellulose has also proved useful due to its slower degradation kinetics in soil and the high stability of the inoculum at room temperature for a long period [70].

Lyophilized microbial cultures can be incorporated into a solid carrier or utilized directly.

The addition in the formulation of carbon sources (e.g., skimmed milk) [71] or stimulatory compounds (e.g., substances present in soil organic matter) that could increase the efficiency of inoculation is also another issue that could be considered while designing a formulation. G. intraradices hyphal growth and root inoculation were increased by specific organic matter components (humic acids or fractions fraction enriched in structures chemically related to 3, 4, 5-trihydroxybenzoic acid or syringic acid) [72, 73]. Treating seeds with a formulation of B. subtilis AF 1 in peat supplemented with chitin or chitin-containing materials showed better control of different soil-borne pathogens and enhanced plant growth than the bacteria culture alone [74].

Natural Carriers

Peat has been commonly used as a carrier for PGPR, particularly for rhizobia inoculants, due to its wide availability and a long history of field trials [75]. When added to peat, PGPR maintain metabolic activity and in some cases can continue to multiply during the storage period, thus increasing their population size, but this can vary with different stains [75]. However, a major drawback of peat is the variability in its quality and composition (acidity), due to its origin from different production sites, which can affect PGPM viability [76]; furthermore, peat holds a large load of microorganisms, which can reduce the shelf life of the inoculant [77]. Similar disadvantages can be listed for carriers made of plant waste materials, which are generally not used for commercial preparations. On the other hand, composts could also be considered as possible carriers, especially when the process of their production involves the use of specific selected strains. For example, adding N-fixing and P-solubilizing bacteria to a vermicompost increased the amount of N and phosphorus availability in the final product [23, 78]. However, composted organic materials are not always useful as AMF carriers. Cellulose-rich amendments could reduce the mycorrhization rate in the case of not fully composted materials [79, 80] even though cellulose can increase the asymbiotic hyphal growth of AMF [81]. Sawdust was shown to be useful as a carrier for production of inocula containing different strains of bacteria [82].

The increasing availability of sludge wastewater has led to consider also this material as a growth medium and carrier for PGPM inoculants. Sludge with heavy metals content below the legal limits was safely used in production of bacterial inoculants [22]. Sludge-based carrier maintained desired rhizobia populations (10^7-10^8 cells g^{-1}), with pH around neutral and an acceptable water holding capacity, after 130 days of storage at 25°C [83].

Coal, clays, and inorganic soils (i.e., lapillus, volcanic pumice or diatomite earths) are available in different regions and can be used as carriers [75]. Their microbial load depends on the site of production (about 10^2-10^3 CFU g^{-1}), but it is generally lower than in organic carriers. Vermiculite, perlite, and bentonite are also available in different countries, but their use is generally limited due to the difficulty in creating a formulation [84]. Indeed, the effect of these carriers on bacteria viability and mobility is dependent on the pH, ion strength, and the electrolyte in solution [85]. Expanded clay has been tested as a carrier for AMF [86], and mycorrhized roots mixed with soil are also the simplest AMF inocula. Among other inorganic compounds, glass beads have also been proposed for AMF inocula [87]. A mixture of organic and inorganic materials have been proved successful in increasing activity and shelf life of Burkholderia sp [88].

All of the above mentioned carriers rely on the absorption of the microorganisms by the substance/matrix of the carrier. This method of inclusion has some drawbacks, particularly in relation to the survival of the microorganisms and their protection during transport, storage, and handling. Nevertheless, some processes with different carriers using such approach have been patented:

- the Belgian patent no. 521.850 for use of diatomaceous earth and colloidal silica for Rhizobium,
- the British patent no. 1.777.077 for the use of bentonite for Rhizobium,
- French Patent no. 1.180.000 using a must juice, to which substances with an adsorbing action are added, such as cellulose, bone meal, kaolin, or silica gel, in the manufacture of preparations rich in bacteria of the Azotobacter group,
- United States Patent no. 4956295 for the stabilization of dried bacteria extended in particulate carriers, where dried viable bacteria are mixed in a particulate carrier composed primarily of

an inorganic salt of low moisture absorbing capacity together with a minor proportion of a silica gel absorbent. The inorganic salts may be sodium or calcium carbonates, bicarbonates, sulfates, or phosphates.

Polymer-Based C5arriers

The increased inte0rest in the application of bacterial preparations as plant protection products has promoted studies aiming at improving their stability and increasing their shelf life. However, the same approach could be successfully applied to products containing PGPRs and AMF. Among the new materials utilized as carriers for PGPM, organic polymers have been evaluated. These are compounds (e.g., polysaccharides) that in the presence of ions or by changing chemical conditions (e.g., a change in pH of the medium) form cross-links that create a complex structure. The polymers encapsulate, or "immobilize", the microorganisms in the matrix and release them gradually through a degradation process. Polymer formulations offer a long shelf life even at ambient temperature since they provide protection against environmental stresses and a consistent batch quality due to standardized production. Nevertheless, storage at cool temperature (4°C) allows to maintain a longer viability of encapsulated cells [12]. These inoculants can be added or mixed with nutrients to improve the survival of the bacteria upon inoculation.

Alginate, a natural polymer of D-mannuronic acid and L-glucuronic acid, is the most commonly used substance for microbial cell encapsulation. It is derived mainly from brown macroalgae such as Macrocystis pyrifera (kelp), but recently also another macroalga (Sargassum sinicola) has been shown to produce alginate of similar physical characteristics [89]. The reaction between alginate and a multivalent cation (e.g., $Ca2^{+}$) forms a gel consisting of a dense three-dimensional lattice with a typical pore-size range of 0.005 to 0.2 mm in diameter [90]; when the alginate solution is dropped into the cation solution beads are formed. Alginate beads generally have a diameter of 2-3 mm, but microbeads with a size of 50 to 200 μm that can entrap up to 10^8 to 10^9 CFU g^{-1} have also been proposed [91].

Inclusion of bacteria in alginate beads has been utilized for different species, either spore forming and non sporulating [12]. Different AMF structures have also been entrapped into alginate matrixes [92, 93] or in beads formed with different polymers [94]. Spores of mycorrhizal fungi were entrapped in alginate film formed in a PVC-coated fibreglass screen [95], and roots of leek seedlings inoculated with this alginate film containing G. mosseae spores were heavily colonized after few weeks of growth in greenhouse conditions. Similar results were obtained with spores obtained from monoxenic cultures embedded into beads [96]. Inclusion of filamentous fungi such as Aspergillus [97] and Actinomycetes has been also proved possible (Malusá, Trzczyński and Taddei, unpublished observations).

Alginate beads can maintain a sufficient amount of live cells to assure inoculation up to several months [65, 98]. However, improving the viability of inocula is still an issue. To tackle it, several approaches have been tested. Adding nutrients (e.g., skimmed milk) to the inoculum [70] or freeze-drying gel beads in presence of glycerol [99] resulted in a prolongation of beads shelf life. Intraradical structures of G. intraradices embedded in alginate beads were still infective after up to 62 months after storage in plastic vials at 4°C [100]. However, it shall be considered that freeze-drying of alginate beads can result in some collapse of the matrix [101]. Therefore, the addition of fillers (material added to the moulding mixture to reduce cost and/or improve mechanical properties) should be considered when planning this technological process. Adding chitin to the beads [102] helped preserve their porous cellular structure resulting in significantly higher porosity values when compared to starch filled beads [103] and resulted in higher bacterial efficacy when evaluating their effect on plants. Addition of 0.5% kaolin to freeze-dried alginate-glycerol beads significantly increased bacterial survival also under UV light radiation [104].

Reducing the cost of the production process and enhancing the physical characteristics of the beads were also obtained by encapsulation and air-drying of bacteria into a mixture made of alginate (3%), standard starch (44.6%), and modified starch (2.4%) [105]. This process allowed to obtain beads that after drying have a water content of 7%, size of 4 mm, and a mechanical resistance of about 105 Newton (features similar to that of grain seeds). Storage at room temperature or at 4°C did not affect the viability of the encapsulated bacteria, which were able to survive up to six months maintaining a final population size of about 10^8 CFU g^{-1}(corresponding to about 10^5 CFU $bead^{-1}$) [106]. However, with this composition, some problems can arise when standardizing and automating the beads formation due to the viscosity of the mixture and the need of a continuous agitation of the stock medium (Malusá and Wawrzyńczak, unpublished observations). Recently, a process using starch industry wastewater as a carbon source for the production of Sinorhizobium meliloti with simultaneous formulation using alginate and soy oil as emulsifier has been proposed, showing a cell viability of more than 10^9 CFU mL^{-1} after 9 weeks of storage [106]. Addition of synthetic zeolite to the alginate mixture did not improve the survival of the embedded microbial cells, nor the physical structure of the beads [65].

Other polymers have been tested with AMF. Carragenan was used to encapsulate AMF structures while hydroxyethylcellulose was used as a gel carrier [107]. Two patents have also been registered:(i)French Patent application no. 77.10254 (corresponding to U.S. Patent no. 4.155.737) which makes use of a polymer gel based on polyacrylamide gel or a silica gel for different microorganisms,(ii)the US patent 5021350 on the process for inclusion of mycorrhizae and actinorhizae in a polymer gel matrix based on at least one polymer from the polysaccharide group, with at least partial cross-linking of the polymer.

Promising New Technologies

Water-in-oil emulsions appear to be a good, yet underutilized, method for storing and delivering microorganisms through liquid formulations [108]. The oil traps the water around the organism and, therefore, slows down water evaporation once applied. This is particularly beneficial for organisms that are sensitive to desiccation or in case of the use for horticultural crops where irrigation systems are in place. Water-in-oil emulsions allow the addition of substances to the oil and/or aqueous phases which could improve both cell viability and release kinetics. However, cell sedimentation during storage is a major issue to be considered. Studies are carried out aiming at solving this problem with the help of nanomaterials. Thickening the oil phase using hydrophobic silica nanoparticles significantly reduced cell sedimentation and improved cell viability during storage [109].

Recently, a new process based on the application of supercritical fluid properties has been tested to encapsulate virus formulations [110] and could also be applied to prepare bacterial inocula. The process, named PGSS (Particles from Gas Saturated Solutions), is carried out at low temperatures and uses carbon dioxide as a supercritical fluid. Therefore, there should be no negative effects on the microorganisms' viability, and the cost of production would be relatively cheap. The final product of the process is almost spherical particles that form a free-flowing powder which can be suspended in water [111]. The possibilities of the PGSS process have already successfully been demonstrated for several solids and liquids [111, 112].

Another interesting new technology is proposing the exploitation of the natural production of bacterial biofilms as a possible carrier, and not only for the production of the inoculum, of defined bacterial or fungi-bacteria consortia. Biofilm production is already obtained for different industrial applications (e.g., wastewater treatment, production of chemical compounds) [113]. Two types of biofilms are employed in that case: biofilms growing onto inert supports (charcoal, resin, concrete, clay brick, sand particles) and biofilms that are formed as a result of aggregate formation. In the first case, biofilms grow all around the particles, and the size of the biofilm particles grows with time usually to several mm in diameter. Biofilm formed by aggregation is called granular biofilm; granule formation may take from several weeks to several months [113].

There are four stages to the development of a mature biofilm: initial attachment, irreversible attachment by the production of EPS, early development, and maturation of biofilm architecture [114]. Particularly critical is the production of EPS, which serves to bind the cell to the surface and to protect it from the surrounding environment. EPS can be composed of polysaccharides, proteins, nucleic acids, or phospholipids. A common EPS produced by bacterial cells in biofilms is the exopolysaccharide alginate [115].

The speed of biofilms formation and maturation is affected by surface, cellular, and environmental factors. Rough surfaces, porous, and less hydrophobic materials tend to enhance biofilm formation [116]. Biofilms tend to form more readily in the presence of optimum nutrients availability, particularly of phosphorous which increases the adhesion ability of cells [117]. High temperature increases the rate of cell growth, EPS production, and surface adhesion, all of which enhance biofilm formation [118]. Biofilm reactors can be assembled in a number of configurations including batch, continuous stirred tank, packed bed, trickling bed, fluidized bed, airlift reactors, upflow anaerobic sludge blanket, and expanded bed reactors [113].

Beneficial biofilms developed in in vitro cultures containing both fungal and bacterial strains were used as biofertilizers for nonlegume species with good efficacy results [24]. Application of a biofilmed inoculant containing a fungal-rhizobia consortium significantly increased N_2fixation in soybean compared to a traditional rhizobium inoculant [119]. Wheat seedlings inoculated with biofilm-producing bacteria exhibited an increased yield in moderate saline soils [120]. Biofilms seem also to help the microorganisms to survive after inoculation even under stress conditions: this is a key aspect for the effectiveness of PGPM inoculation under agricultural conditions. Inocula made with biofilms were shown to allow their rhizobia survive at high salinity (400 mM NaCl) by 105-fold compared to rhizobial monocultures [24]. Interestingly, beneficial endophytes were observed to produce higher acidity and plant growth-promoting hormones than their mono- or mixed cultures with no biofilm formation [121].

Technologies used for the production of living hybrids materials could be a new frontier in the development of carriers for PGPMs. Silica has appeared as a promising host for microorganisms encapsulation: immobilization pathways are based on immobilization of population bacteria dispersed into a silica gel. Bacteria can be either entrapped into alginate microbeads coated with silica membranes or into macrocavities created inside the silica matrix. Such material improves the mechanical properties of the alginate bead, reduces cell leakage, and enhances cell viability [122].

The application of bionanotechnologies could also provide new avenues for the development of carrier-based microbial inocula [123, 124]. Nanotechnology employs nanoparticles which are made of inorganic or organic materials that are defined by having one or more dimensions in the order of 100 nm or less [125]. The integration of whole cells with nanostructures leads to hybrid systems that have numerous applications in many fields including agriculture [126]. Indeed, even though nanoscale constructs are smaller than cells, macroscopic filters, made of radially aligned carbon nanotube walls, able to absorb Escherichia coli, were fabricated [127]. The same technology could therefore be applied to collect bacterial cells from fermentation processes and deliver them to the plant. The physical stability and the high surface area of nanotubes, together with the ease and cost-effective fabrication of nanotube membranes may thus expand their use in the production of biofertilizer.

The use of nanoformulations may enhance the stability of biofertilizers and biostimulators with respect to dessication, heat, and UV inactivation. The addition of hydrophobic silica nanoparticles of 7–14 nm to the water-in-oil emulsion formulation of the biopesticide fungusLagenidium giganteum reduced the desiccation of the mycelium. The physical features of the formulation were improved and the microorganism was still effective after 12 weeks of storage at room temperature [109].

APPLICATION METHODS

A limited array of methods exists for the delivery of PGPM to crops in the field. Farmers are not keen on purchasing specialized equipment to be used for microbial-based products. Therefore, formulated inocula should be readily applied using standard farming machinery and straightforward methods. Inoculation can be done through application to the plant material or to the soil. The latter method can be more convenient for the farmer because of less time required, but generally a higher amount of inoculant is then needed. Soil inoculation can be done either with solid or liquid formulations. Normally, the inert material is mixed with the inoculum in the factory, but it could be mixed by the farmer prior to application, especially when liquid formulations are used. The use of fertilizers which were produced mixing organic matrixes and insoluble phosphates with addition of selected P-solubilizing microorganisms can also be considered a method to apply PGPM to crops: it increases the availability of nutrients (particularly of P) to plants and eventually affects the tolerance of the plant to soil pathogens [128].

Application methods depend on the kind of crop concerned: annual crops can be inoculated by broadcasting the inoculum over the soil surface, alone or together with seeds, or by in-furrow application, seed dressing, or coating; tree crops can be initially inoculated by root dipping or seedling inoculation [64]. Application to already established orchards or plantations can present some technical problems, when inocula have to be distributed to the soil [129]. The need to deliver the PGPM as close as possible to the root system can be fulfilled by liquid formulations applied through a fertigation system. Trials using alginate beads and polyurethane foam as carriers to deliver PGPM into the water solution are showing the feasibility of this approach (Malusá, Trzcinski and Treder, unpublished observations); however, there is the need to either soak the foam for some time into the water tank or dissolve the beads using a citrate solution. Alternatively, powder materials can be buried near the roots using a harrow-like device associated with a distributor. Subsequent irrigation would increase the transfer of the inocula toward the roots and may enhance the efficacy of the inoculation, creating better moisture conditions which favor the movement of the bacteria in soil [130].

Microbial populations in the soil could dilute or counteract the effect of introduced PGPM. In the case of PGPR, the recovery of the inoculated strains in the soil or on root rhizosphere was limited to 30–40 days after inoculation [66]. Therefore, repeated applications (3-4) during the growing season, with an interval of 2–4 weeks, increase the effectiveness of PGPM applications.

CONCLUSIONS

A better understanding of the different conditions and features of the interrelationships in the soil-plant-microorganism system is needed to improve the efficacy of PGPM inocula applications in the field. Indeed, particularly for bacteria, many factors are involved in determining their rhizocompetence. Together with the genotype and physiological state of the inoculated strain, the size and composition of the populations sustained by the rhizosphere is determined by several environmental factors: soil pH, mineral nutrients, and water content; species, genotype, and physiological state of the plant; the presence of other microbial species [131].

Several isolates have been obtained in the last decades showing plant growth enhancement or biocontrol properties. Yet the knowledge of PGPM behaviour at the root level and their function in the field environment is still limited [36, 37].

Considering the beneficial effects of PGPR and AMF, studies using inoculant mixtures are opening a new approach to the subject.

These studies would facilitate the designing of large consortia of inocula that bring about a synergistic promotion of plant growth or have multitasking features [132]. However, such mixtures are more technically demanding, since they increase the difficulties in designing a proper inoculant that would fit the different strains and kinds of microorganisms. The designing of biofilmed-based carriers [24] or of encapsulation techniques which allow the production of macrocapsules consisting of a core and an envelope could facilitate the development of biofertilizers formed of microbial consortia.

Most inoculants selected to date have been designed for annual crops (mainly legumes, cereals, and some vegetables). However, there is an increasing demand from other agricultural sectors such as fruit and vegetable production, and particularly from organic farming and integrated production systems, where synthetic inputs are not allowed or their use is limited by legal restrictions. Soilless and protected crops can also be an interesting market for commercial inoculants, where the predictability of PGPM applications should be higher than in open fields due to the use of inert substrates and controlled growth conditions. The selection of specific strains for all these crops can further expand the market for inocula and support the shift in agriculture toward more sustainable production systems. However, also the development of technologies aiming at an efficient delivery of the inocula, probably by modification of sprayers and sprinklers normally used for plant protection or irrigation, could further increase the use and enhance the reliability of PGPM applications under agricultural conditions.

The future challenges in selecting PGPM are related to the attempts at alleviating abiotic stress conditions in crops (i.e., drought, salinity, inorganic, and organic pollutants) and improving food quality. Improvements in the production process for consortia of microbial inocula, the development of new carriers based on nanoparticles, optimization of application devices and of the time of application for polyannual crops, are all issues requiring further research to widen the implementation and efficient use of PGPM in agriculture. Such need is also prompted by the policy decisions supporting sustainable practices, as well as by the reassessment of the safety of plant protection agents, currently underway both in EU and USA, which can further foster the market potential for PGPM.

ACKNOWLEDGMENTS

The work has been supported by a grant from the EU Regional Development Fund through the Polish Innovation Economy Operational Program, Contract no. UDA-POIG.01.03.01-10-109/08-00.

REFERENCES

1. W. Khan, U. P. Rayirath, S. Subramanian et al., "Seaweed extracts as biostimulants of plant growth and development," Journal of Plant Growth Regulation, vol. 28, no. 4, pp. 386–399, 2009.
2. A. Gousterova, M. Nustorova, P. Christov, P. Nedkov, G. Neshev, and E. Vasileva-Tonkova, "Development of a biotechnological procedure for treatment of animal wastes to obtain inexpensive biofertilizer," World Journal of Microbiology and Biotechnology, vol. 24, no. 11, pp. 2647–2652, 2008
3. P. Jeffries, S. Gianinazzi, S. Perotto, K. Turnau, and J. M. Barea, "The contribution of arbuscular mycorrhizal fungi in sustainable maintenance of plant health and soil fertility," Biology and Fertility of Soils, vol. 37, no. 1, pp. 1–16, 2003.
4. A. R. Podile and G. K. Kishore, "Plant growth-promoting rhizobacteria," in Plant-Associated Bacteria, S. S. Gnanamanickam, Ed., pp. 195–230, Springer, Amsterdam, The Netherlands, 2006.
5. C. Franche, K. Lindström, and C. Elmerich, "Nitrogen-fixing bacteria associated with leguminous and non-leguminous plants," Plant and Soil, vol. 321, no. 1-2, pp. 35–59, 2009
6. J. K. Vessey, "Plant growth promoting rhizobacteria as biofertilizers," Plant and Soil, vol. 255, no. 2, pp. 571–586, 2003
7. B. J. J. Lugtenberg, T. F. C. Chin-A-Woeng, and G. V. Bloemberg, "Microbe-plant interactions: principles and mechanisms," Antonie Van Leeuwenhoek, vol. 81, no. 1–4, pp. 373–383, 2002
8. A. O. Adesemoye, H. A. Torbert, and J. W. Kloepper, "Enhanced plant nutrient use efficiency with PGPR and AMF in an integrated nutrient management system,"Canadian Journal of Microbiology, vol. 54, no. 10, pp. 876–886, 2008.

9. E. Malusà, L. Sas-Paszt, and E. Zurawicz, "The effect of a mycorrhiza-bacteria substrate and foliar fertilization on growth response and rhizosphere pH of three strawberry cultivars," International Journal of Fruit Science, vol. 6, no. 4, pp. 25–41, 2007
10. R. J. Lempert, P. Norling, C. G. Pernin, S. A. Resetar, and S. Mahnovski, Next Generation Environmental Technologies: Benefits and Barriers, Rand, New York, NY, USA, 2003.
11. T. J. Avis, V. Gravel, H. Antoun, and R. J. Tweddell, "Multifaceted beneficial effects of rhizosphere microorganisms on plant health and productivity," Soil Biology and Biochemistry, vol. 40, no. 7, pp. 1733–1740, 2008
12. Y. Bashan, "Inoculants of plant growth-promoting bacteria for use in agriculture,"Biotechnology Advances, vol. 16, no. 4, pp. 729–770, 1998
13. R. T. Koide and B. Mosse, "A history of research on arbuscular mycorrhiza,"Mycorrhiza, vol. 14, no. 3, pp. 145–163, 2004
14. J. W. Kloepper, C.-M. Ryu, and S. Zhang, "Induced systemic resistance and promotion of plant growth by Bacillus spp," Phytopathology, vol. 94, no. 11, pp. 1259–1266, 2004
15. M. Lucy, E. Reed, and B. R. Glick, "Applications of free living plant growth-promoting rhizobacteria," Antonie Van Leeuwenhoek, vol. 86, no. 1, pp. 1–25, 2004
16. Y. Okon and C. A. Labandera-Gonzalez, "Agronomic applications of Azospirillum an evaluation of 20 years worldwide field inoculation," Soil Biology and Biochemistry, vol. 26, no. 12, pp. 1591–1601, 1994
17. D. R. Fravel, "Commercialization and implementation of biocontrol," Annual Review of Phytopathology, vol. 43, pp. 337–359, 2005
18. V. Artursson, R. D. Finlay, and J. K. Jansson, "Interactions between arbuscular mycorrhizal fungi and bacteria and their potential for stimulating plant growth,"Environmental Microbiology, vol. 8, no. 1, pp. 1–10, 2006
19. M. Guillon, "Current world situation on acceptance and marketing of biological control agents (BCAS). Position Paper by the President of IBMA, International Biocontrol Manufacturers Association," http://www.ibma.ch/papers.html, 2006.

20. F. R. Schmidt, "Optimization and scale up of industrial fermentation processes," Applied Microbiology and Biotechnology, vol. 68, no. 4, pp. 425–435, 2005
21. A. S. Vidyarthi, M. Desrosiers, R. D. Tyagi, and J. R. Valéro, "Foam control in biopesticide production from sewage sludge," Journal of Industrial Microbiology and Biotechnology, vol. 25, no. 2, pp. 86–92, 2000.
22. F. B. Rebah, R. D. Tyagi, D. Prévost, and R. Y. Surampalli, "Wastewater sludge as a new medium for rhizobial growth," Water Quality Research Journal of Canada, vol. 37, no. 2, pp. 353–370, 2002
23. M. Vassileva, M. Serrano, V. Bravo et al., "Multifunctional properties of phosphate-solubilizing microorganisms grown on agro-industrial wastes in fermentation and soil conditions," Applied Microbiology and Biotechnology, vol. 85, no. 5, pp. 1287–1299, 2010
24. G. Seneviratne, J. S. Zavahir, W. M. M. S. Bandara, and M. L. M. A. W. Weerasekara, "Fungal-bacterial biofilms: their development for novel biotechnological applications,"World Journal of Microbiology and Biotechnology, vol. 24, no. 6, pp. 739–743, 2008
25. S. Ude, D. L. Arnold, C. D. Moon, T. Timms-Wilson, and A. J. Spiers, "Biofilm formation and cellulose expression among diverse environmental Pseudomonas isolates," Environmental Microbiology, vol. 8, no. 11, pp. 1997–2011, 2006.
26. L. L. Hung and D. M. Sylvia, "Production of vesicular-arbuscular mycorrhizal fungus inoculum in aeroponic culture," Applied and Environmental Microbiology, vol. 54, pp. 353–357, 1992.
27. D.-G. Strullu and C. Romand, "Méthode d›obtention d›endomycorhizes à vésicules et arbuscules en conditions axéniques," Comptes Rendus de l›Académie des Sciences, vol. 303, pp. 245–250, 1986.
28. D.-G. Strullu, C. Romand, and C. Plenchette, "Axenic culture and encapsulation of the intraradical forms of Glomus spp," World Journal of Microbiology and Biotechnology, vol. 7, no. 3, pp. 292–297, 1991

29. M. St-Arnaud, C. Hamel, B. Vimard, M. Caron, and J. A. Fortin, "Enhanced hyphal growth and spore production of the arbuscular mycorrhizal fungus Glomus intraradicesin an in vitro system in the absence of host roots," Mycological Research, vol. 100, no. 3, pp. 328–332, 1996.

30. D. D. Douds Jr., "Increased spore production by Glomus intraradices in the split-plate monoxenic culture system by repeated harvest, gel replacement, and resupply of glucose to the mycorrhiza," Mycorrhiza, vol. 12, no. 4, pp. 163–167, 2002

31. A. Verma and A. Aldholeya, "Cost-economics of existing methodologies for inoculum production of vesicular—arbuscular mycorrhizal fungi," in Concepts in Mycorrhizal Research, K. G. Mukerji, Ed., pp. 179–194, Kluwer Academic, Dodrecht, The Netherlands, 1996.

32. A. Adholeya, P. Tiwari, and R. Singh, "Large-scale inoculum production of arbuscular mycorrhizal fungi on root organs and inoculation strategies," in Soil Biology, S. Declerck, D. G. Strullu, and A. Fortin, Eds., vol. 4 of In Vitro Culture of Mycorrhizas, pp. 315–140, Springer, Heidelberg, Germany, 2005.

33. M. IJdo, S. Cranenbrouck, and S. Declerck, "Methods for large-scale production of AM fungi: past, present, and future," Mycorrhiza, vol. 21, no. 1, pp. 1–16, 2011

34. H. Stockinger, M. Krüger, and A. Schüßler, "DNA barcoding of arbuscular mycorrhizal fungi," New Phytologist, vol. 187, no. 2, pp. 461–474, 2010

35. F. Feldmann and C. Grotkass, "Directed inoculum production—shall we be able to design AMF populations to achieve predictable symbiotic effectiveness?" in Mycorrhizal Technology in Agriculture: From Genes to Bioproducts, S. Gianinazzi, H. Schüepp, J. M. Barea, and K. Haselwandter, Eds., pp. 261–279, Birkhäuser, Basel, Switzerland, 2002.

36. G. Berg, "Plant-microbe interactions promoting plant growth and health: perspectives for controlled use of microorganisms in agriculture," Applied Microbiology and Biotechnology, vol. 84, no. 1, pp. 11–18, 2009

37. P. Frey-Klett, J. Garbaye, and M. Tarkka, "The mycorrhiza helper bacteria revisited,"New Phytologist, vol. 176, no. 1, pp. 22–36, 2007

38. J. F. Toljander, V. Artursson, L. R. Paul, J. K. Jansson, and R. D. Finlay, "Attachment of different soil bacteria to arbuscular mycorrhizal fungal extraradical hyphae is determined by hyphal vitality and fungal species," FEMS Microbiology Letters, vol. 254, no. 1, pp. 34–40, 2006
39. A. Gange, "Arbuscular mycorrhizal fungi, Collembola and plant growth," Trends in Ecology and Evolution, vol. 15, no. 9, pp. 369–372, 2000
40. Y. Bashan, G. Holguin, and L. E. de-Bashan, "Azospirillum-plant relationships: physiological, molecular, agricultural, and environmental advances (1997-2003),"Canadian Journal of Microbiology, vol. 50, no. 8, pp. 521–577, 2004
41. A. A. Belimov, A. P. Kojemiakov, and C. V. Chuvarliyeva, "Interaction between barley and mixed cultures of nitrogen fixing and phosphate-solubilizing bacteria," Plant and Soil, vol. 173, no. 1, pp. 2937–2942, 1995
42. J. M. Barea, R. Azcón, and C. Azcón-Aguilar, "Mycorrhizosphere interactions to improve plant fitness and soil quality," Antonie Van Leeuwenhoek, vol. 81, no. 1–4, pp. 343–351, 2002
43. J. W. Kloepper, "Host specificity in microbe-microbe interactions," BioScience, vol. 46, no. 6, pp. 406–409, 1996.
44. N. Vassilev, M. Vassileva, R. Azcon, and A. Medina, "Preparation of gel-entrapped mycorrhizal inoculum in the presence or absence of Yarowia lipolytica," Biotechnology Letters, vol. 23, no. 11, pp. 907–909, 2001. · ·
45. B. Biró, K. Köves-Péchy, I. Vörös, T. Takács, P. Eggenberger, and R. J. Strasser, "Interrelations between Azospirillum and Rhizobium nitrogen fixers and arbuscular mycorrhizal fungi in the rhizosphere of alfalfa in sterile AMF free or normal conditions," Applied Soil Ecology, vol. 15, no. 2, pp. 159–168, 2000. · ·
46. J. Secilia and D. J. Bagyaraj, "Bacteria and actinomycetes associated with pot cultures of vesiculararbuscular mycorrhizas," Canadian Journal of Microbiology, vol. 33, pp. 1069–1073, 1987.
47. C. Azcón-Aguilar and J. M. Barea, "Applying mycorrhiza biotechnology to horticulture: significance and potentials," Scientia Horticulturae, vol. 68, no. 1–4, pp. 1–24, 1997. · ·

48. P. E. Lovato, H. Schüepp, A. Trouvelot, and S. Gianinazzi, "Application of arbuscular mycorrhizal fungi (AMF) in orchard and ornamental plants," in Mycorrhiza Structure, Function, Molecular Biology and Biotechnology, A. Varma and B. Hock, Eds., pp. 521–559, Springer, Heidelberg, Germany, 1995.

49. J. M. Barea, R. Azcón, and C. Azcón-Aguilar, "Interactions between mycorrhizal fungi and bacteria to improve plant nutrient cycling and soil structure," in Microorganisms in Soils: Roles in Genesis and Functions, F. Buscot and A. Varma, Eds., pp. 195–212, Springer, Berlin, Germany, 2005.

50. E. Gamalero, A. Trotta, N. Massa, A. Copetta, M. G. Martinotti, and G. Berta, "Impact of two fluorescent pseudomonads and an arbuscular mycorrhizal fungus on tomato plant growth, root architecture and P acquisition," Mycorrhiza, vol. 14, no. 3, pp. 185–192, 2004.

51. M. Toro, R. Azcón, and J. M. Barea, "Improvement of arbuscular mycorrhizal development by inoculation with phosphate-solubilizing rhizobacteria to improve rock phosphate bioavailability (P^{32}) and nutrient cycling," Applied and Environmental Microbiology, vol. 63, pp. 4408–4412, 1997.

52. N. Vassilev, M. Vassileva, R. Azcon, and A. Medina, "Interactions of an arbuscular mycorrhizal fungus with free or co-encapsulated cells of Rhizobium trifoli and Yarowia lipolytica inoculated into a soil-plant system," Biotechnology Letters, vol. 23, no. 2, pp. 149–151, 2001.

53. J. Lisette, C. Xavier, and J. J. Germida, "Selective interactions between arbuscular mycorrhizal fungi and Rhizobium leguminosarum bv. viceae enhance pea yield and nutrition," Biology and Fertility of Soils, vol. 37, no. 5, pp. 261–267, 2003.

54. J. M. Barea, M. J. Pozo, R. Azcón, and C. Azcón-Aguilar, "Microbial co-operation in the rhizosphere," Journal of Experimental Botany, vol. 56, no. 417, pp. 1761–1778, 2005.

55. M. Gryndler, M. Vosátka, H. Hršelová, V. Catská, I. Chvátalová, and J. Jansa, "Effect of dual inoculation with arbuscular mycorrhizal fungi and bacteria on growth and mineral nutrition of strawberry," Journal of Plant Nutrition, vol. 25, no. 6, pp. 1342–1358, 2002.

56. A. Medina, A. Probanza, F. J. Gutierrez Mañero, and R. Azcón, "Interactions of arbuscular-mycorrhizal fungi and Bacillus strains and their effects on plant growth, microbial rhizosphere activity (thymidine and leucine incorporation) and fungal biomass (ergosterol and chitin)," Applied Soil Ecology, vol. 22, no. 1, pp. 15–28, 2003.
57. E. Malusà, G. Buffa, and J. Ciesielska, "Effect of different fertilisation management on photosynthesis, yield and fruit quality of peach," in Plant Nutrition. Food Security and Sustainability of Agro-Ecosystems through Basic and Applied Research, W. Horst, Ed., pp. 332–333, Kluwer Academic, London, UK, 2001.
58. E. Malusà, E. Laurenti, E. Ghibaudi, and L. Rolle, "Influence of organic and conventional management on yield and composition of grape cv. ‹Grignolino›," Acta Horticulturae, vol. 640, pp. 135–141, 2004.
59. E. Malusà, L. Sas-Paszt, and J. Ciesielska, "Effect of new organic fertilizers on growth of strawberry cv. Elsanta. Preliminary results," in Proceedings of the XIV International Conference on Organic Fruit Growing, pp. 361–365, Foerdergemeinschaft Oekologischer Obstbau e.V., Hohenheim, Germany, 2010.
60. L. Sas-Paszt, E. Żurawicz, A. Masny, et al., "The use of biostimulators in small fruit growing," in Biostimulators in Modern Agriculture. Fruit Crops, A. Sadowski, Ed., pp. 76–90, Wieś Jutra, Warsaw, Poland, 2008.
61. R. S. Smith, "Legume inoculant formulation and application," Canadian Journal of Microbiology, vol. 38, no. 6, pp. 485–492, 1992.
62. H. H. Keyser, P. Somasegaran, and B. B. Bohlool, "Rhizobial ecology and technology," inSoil Microbial Ecology: Applications in Agricultural and Environmental Management, E. B. Metting, Ed., pp. 205–226, Marcel Dekker, New York, NY, USA, 1993.
63. S. V. Hegde and G. P. Brahmaprakash, "A dry granular inoculant of Rhizobium for soil application," Plant and Soil, vol. 144, no. 2, pp. 309–311, 1992.
64. R. Muresu, L. Sulas, and S. Caredda, "Legume—Rhizobium symbiosis: characteristics and prospects of inoculation," Rivoluzione Agronomica, vol. 37, pp. 33–45, 2003.

65. P. Trzciński, E. Malusá, and L. Sas Paszt, "Survival of PGPR in beads of biodegrdable polimer," in Proceedings of the Ogólnopolską Naukową Konferencję Ekologiczną „Osiągnięcia i Możliwości Rozwoju Badań i Wdrożeń w Ekologicznej Produkcji Ogrodniczej, pp. 181–182, Skierniewice, Poland, 2011.

66. Y. Bashan, M. E. Puente, M. N. Rodriguez-Mendoza et al., "Survival of Azospirillum brasilense in the bulk soil and rhizosphere of 23 soil types," Applied and Environmental Microbiology, vol. 61, no. 5, pp. 1938–1945, 1995.

67. M. Bonkowski, "Protozoa and plant growth: the microbial loop in soil revisited," New Phytologist, vol. 162, no. 3, pp. 617–631, 2004.

68. G. Ciafardini and C. Barbieri, "Effects of cover inoculation of soybean on nodulation, nitrogen fixation and yield," Agronomy Journal, vol. 79, pp. 645–648, 1987.

69. J. Crowe, J. Carpenter, and L. Crowe, "The role of vitrification in anhydrobiosis,"Annual Review of Physiology, vol. 60, pp. 73–103, 1998.

70. A. Hernández, F. Weekers, J. Mena, C. Borroto, and P. Thonart, "Freeze-drying of the biocontrol agent Tsukamurlla paurometabola C-924: predicted stability of formulated powders," Industrial Biotechnology, vol. 2, no. 3, pp. 209–212, 2006.

71. P. Trivedi, A. Pandey, and L. M. S. Palni, "Carrier-based preparations of plant growth-promoting bacterial inoculants suitable for use in cooler regions," World Journal of Microbiology and Biotechnology, vol. 21, no. 6-7, pp. 941–945, 2005.

72. M. Gryndler, H. Hršelová, R. Sudová, H. Gryndlerová, V. Řezáčová, and V. Merhautová, "Hyphal growth and mycorrhiza formation by the arbuscular mycorrhizal fungus Glomus claroideum BEG 23 is stimulated by humic substances," Mycorrhiza, vol. 15, no. 7, pp. 483–488, 2005.

73. M. Gryndler, H. Hršelová, T. Cajthaml et al., "Influence of soil organic matter decomposition on arbuscular mycorrhizal fungi in terms of asymbiotic hyphal growth and root colonization," Mycorrhiza, vol. 19, no. 4, pp. 255–266, 2009.

74. K. Manjula and A. R. Podile, "Chitin-supplemented formulations improve biocontrol and plant growth promoting efficiency of Bacillus subtilis AF 1," Canadian Journal of Microbiology, vol. 47, no. 7, pp. 618–625, 2001.
75. R. S. Smith, "Inoculant formulations and applications to meet changing needs," inNitrogen Fixation: Fundamentals and Applications, I. A. Tikhonovich, N. A. Provorov, V. I. Romanov, and W. E. Newton, Eds., pp. 653–657, Kluwer Academic, Dodrecht, The Netherlands, 1995.
76. V. Corich, E. Bosco, A. Giacomini, M. Basaglia, A. Squartini, and M. P. Nuti, "Fate of genetically modified Rhizobium leguminosarum biovar viciae during long-term storage of commercial inoculants," Journal of Applied Bacteriology, vol. 81, no. 3, pp. 319–328, 1996.
77. J. D. van Elsas and C. E. Heijnen, "Methods for the introduction of bacteria into soil: a review," Biology and Fertility of Soils, vol. 10, no. 2, pp. 127–133, 1990.
78. V. Kumar and K. P. Singh, "Enriching vermicompost by nitrogen fixing and phosphate solubilizing bacteria," Bioresource Technology, vol. 76, no. 2, pp. 173–175, 2001.
79. L. Avio and M. Giovannetti, "Vesicular-arbuscular mycorrhizal infection of lucerne roots in a cellulose-amended soil," Plant and Soil, vol. 112, no. 1, pp. 99–104, 1988.
80. S. Ravnskov, J. Larsen, P. A. Olsson, and I. Jakobsen, "Effects of various organic compounds on growth and phosphorus uptake of an arbuscular mycorrhizal fungus,"New Phytologist, vol. 141, no. 3, pp. 517–524, 1999.
81. M. Gryndler, M. Vosátka, H. Hršelová, I. Chvátalová, and J. Jansa, "Interaction between arbuscular mycorrhizal fungi and cellulose in growth substrate," Applied Soil Ecology, vol. 19, no. 3, pp. 279–288, 2002.
82. N. K. Arora, E. Khare, R. Naraian, and D. K. Maheshwari, "Sawdust as a superior carrier for production of multipurpose bioinoculant using plant growth promoting rhizobial and pseudomonad strains and their impact on productivity of Trifolium repense,"Current Science, vol. 95, no. 1, pp. 90–94, 2008.

83. F. Ben Rebah, R. D. Tyagi, and D. Prévost, "Wastewater sludge as a substrate for growth and carrier for rhizobia: the effect of storage conditions on survival of Sinorhizobium meliloti," Bioresource Technology, vol. 83, no. 2, pp. 145–151, 2002.
84. P. D. Millner and D. G. Kitt, "The Beltsville method for soilless production of vesicular-arbuscular mycorrhizal fungi," Mycorrhiza, vol. 2, no. 1, pp. 9–15, 1992.
85. N. Lahav, "Adsorption of sodium bentonite particles on Bacillus subtilis," Plant and Soil, vol. 17, no. 2, pp. 191–208, 1962.
86. C. Plenchette, V. Furlan, and J. A. Fortin, "Responses of endomycorrhizal plants grown in calcined montmorillonite clay to different levels of phosphorus. I. Effect on growth and mycorrhizal development," Canadian Journal of Botany, vol. 61, pp. 1377–1383, 1983.
87. D. Redecker, H. Thierfelder, and D. Werner, "A new cultivation system for arbuscular mycorrhizal fungi on glass beads," Journal of Applied Botany: Angewandte Botanik, vol. 69, pp. 183–188, 1995.
88. R. Anandham, K. H. Choi, P. I. Gandhi, et al., "Evaluation of shelf life and rock phosphate solubilisation of Burkholderia sp. in nutrient-amended clay, rice bran and rock phosphate-based granular formulation," World Journal of Microbiology and Biotechnology, vol. 23, pp. 1121–1129, 2007.
89. R. Yabur, Y. Bashan, and G. Hernández-Carmona, "Alginate from the macroalgaeSargassum sinicola as a novel source for microbial immobilization material in wastewater treatment and plant growth promotion," Journal of Applied Phycology, vol. 19, no. 1, pp. 43–53, 2007.
90. O. Smidsrod and G. Skjak-Braek, "Alginate as immobilization matrix for cells," Trends in Biotechnology, vol. 8, no. 3, pp. 71–78, 1990.
91. Y. Bashan, J. P. Hernandez, L. A. Leyva, and M. Bacilio, "Alginate microbeads as inoculant carriers for plant growth-promoting bacteria," Biology and Fertility of Soils, vol. 35, no. 5, pp. 359–368, 2002.
92. F. Ganry, H. G. Diem, and Y. R. Dommergues, "Effect of inoculation with Glomus mosseae on nitrogen fixation by field

grown soybeans," Plant and Soil, vol. 68, no. 3, pp. 321–329, 1982.

93. D.-G. Strullu and C. Plenchette, "The entrapment of Glomus sp. in alginate beads and their use as root inoculum," Mycological Research, vol. 95, pp. 1194–1196, 1991.
94. N. Vassilev, I. Nikolaeva, and M. Vassileva, "Polymer-based preparation of soil inoculants: applications to arbuscular mycorrhizal fungi," Re-views in Environmental Science and Biotechnology, vol. 4, no. 4, pp. 235–243, 2005.
95. C. Calvet, A. Campribu, and R. Rodrguez-Kabana, "Inclusion of arbuscular mycorrhizal fungi in alginate films for experimental studies and plant inoculation," HortScience, vol. 31, pp. 285–288, 1996.
96. S. Declerck, D. G. Strullu, and C. Plenchette, "In vitro mass-production of the arbuscular mycorrhizal fungus, Glomus versiforme, associated with Ri T-DNA transformed carrot roots," Mycological Research, vol. 100, no. 10, pp. 1237–1242, 1996.
97. R. Jain, J. Saxena, and V. Sharma, "The evaluation of free and encapsulated Aspergillus awamori for phosphate solubilization in fermentation and soil-plant system," Applied Soil Ecology, vol. 46, no. 1, pp. 90–94, 2010.
98. J. A. van Veen, L. S. van Overbeek, and J. D. van Elsas, "Fate and activity of microorganisms introduced into soil," Microbiology and Molecular Biology Reviews, vol. 61, no. 2, pp. 121–135, 1997.
99. Y. Tal, J. van Rijn, and A. Nussinovitch, "Improvement of structural and mechanical properties of denitrifying alginate beads by freeze-drying," Biotechnology Progress, vol. 13, no. 6, pp. 788–793, 1997.
100. C. Plenchette and D. G. Strullu, "Long-term viability and infectivity of intraradical forms of Glomus intraradices vesicles encapsulated in alginate beads," Mycological Research, vol. 107, no. 5, pp. 614–616, 2003.
101. D. Rassis, A. Nussinovitch, and I. S. Saguy, "Collapse, shrinkage and structural changes in dried alginate gels containing fillers," Food Hydrocolloids, vol. 16, no. 2, pp. 139–151, 2002.

102. C. Zohar-Perez, I. Chet, and A. Nussinovitch, "Mutual relationships between soils and biological carrier systems," Biotechnology and Bioengineering, vol. 92, no. 1, pp. 54–60, 2005.
103. Y. Tal, J. van Rijn, and A. Nussinovitch, "Improvement of mechanical and biological properties of freeze-dried denitrifying alginate beads by using starch as a filler and carbon source," Applied Microbiology and Biotechnology, vol. 51, no. 6, pp. 773–779, 1999.
104. C. Zohar-Perez, L. Chernin, I. Chet, and A. Nussinovitch, "Structure of dried cellular alginate matrix containing fillers provides extra protection for microorganisms against UVC radiation," Radiation Research, vol. 160, no. 2, pp. 198–204, 2003.
105. E. Ivanova, E. Teunou, and D. Poncelet, "Alginate based macrocapsules as inoculants carriers for production of nitrogen biofertilizers," in Proceedings of the Balkan Conference of Biology, B. Gruev, M. Nikolova, and A. Donev, Eds., pp. 90–108, Plovdiv, Bulgaria, 2005.
106. T. Rouissi, R. P. John, S. K. Brar, R. D. Tyagi, and D. Prévost, "Original research: centrifugal recovery of rhizobial cells from fermented starch industry wastewater & development of stable formulation," Industrial Biotechnology, vol. 6, no. 1, pp. 41–49, 2010.
107. D. M. Sylvia and A. G. Jarstfer, "Sheared-root inocula of vesicular-arbuscular mycorrhizal fungi," Applied and Environmental Microbiology, vol. 58, no. 1, pp. 229–232, 1992.
108. J. S. VanderGheynst, H. Scher, and G. Hong-Yun, "Design of formulations for improved biological control agent viability and sequestration during storage," Industrial Biotechnology, vol. 2, no. 3, pp. 213–219, 2006.
109. J. S. Vandergheynst, H. B. Scher, H.-Y. Guo, and D. L. Schultz, "Water-in-oil emulsions that improve the storage and delivery of the biolarvacide Lagenidium giganteum,"BioControl, vol. 52, no. 2, pp. 207–229, 2007.
110. M. Pemsel, S. Schwab, A. Scheurer, D. Freitag, R. Schatz, and E. Schlücker, "Advanced PGSS process for the encapsulation of the biopesticide Cydia pomonella granulovirus,"Journal of Supercritical Fluids, vol. 53, no. 1–3, pp. 174–178, 2010.

111. M. J. Cocero, Á. Martin, F. Mattea, and S. Varona, "Encapsulation and co-precipitation processes with supercritical fluids: fundamentals and applications," Journal of Supercritical Fluids, vol. 47, no. 3, pp. 546–555, 2009.
112. S.-D. Yeo and E. Kiran, "Formation of polymer particles with supercritical fluids: a review," Journal of Supercritical Fluids, vol. 34, no. 3, pp. 287–308, 2005.
113. N. Qureshi, B. A. Annous, T. C. Ezeji, P. Karcher, and I. S. Maddox, "Biofilm reactors for industrial bioconversion process: employing potential of enhanced reaction rates,"Microbial Cell Factories, vol. 4, article 24, 2005.
114. P. Stoodley, K. Sauer, D. G. Davies, and J. W. Costerton, "Biofilms as complex differentiated communities," Annual Review of Microbiology, vol. 56, pp. 187–209, 2002.
115. I. W. Sutherland, "Biofilm exopolysaccharides: a strong and sticky framework,"Microbiology, vol. 147, no. 1, pp. 3–9, 2001.
116. W. G. Characklis, G. A. McFeters, and K. C. Marshall, "Physiological ecology in biofilm systems," in Biofilms, W. G. Characklis and K. C. Marshall, Eds., vol. 37, pp. 67–72, John Wiley & Sons, New York, NY, USA, 1990.
117. J. Büchs, N. Mozes, C. Wandrey, and P. G. Rouxhet, "Cell adsorption control by culture conditions," Applied Microbiology and Biotechnology, vol. 29, no. 2-3, pp. 119–128, 1988.
118. A. P. Annachhatre and S. M. R. Bhamidimarri, "Microbial attachment and growth in fixed-film reactors: process startup considerations," Biotechnology Advances, vol. 10, no. 1, pp. 69–91, 1992.
119. H. S. Jayasinghearachchi and G. Seneviratne, "A bradyrhizobial-Penicillium spp. biofilm with nitrogenase activity improves N2 fixing symbiosis of soybean," Biology and Fertility of Soils, vol. 40, no. 6, pp. 432–434, 2004.
120. M. Ashraf, S. Hasnain, O. Berge, and T. Mahmood, "Inoculating wheat seedlings with exopolysaccharide-producing bacteria restricts sodium uptake and stimulates plant growth under salt stress," Journal of Food Science, vol. 71, no. 3, pp. 89–99, 2004.
121. W. M. M. S. Bandara, G. Seneviratne, and S. A. Kulasooriya, "Interactions among endophytic bacteria and fungi: effects and

potentials," Journal of Biosciences, vol. 31, no. 5, pp. 645–650, 2006.

122. E. Callone, R. Campostrini, G. Carturan, A. Cavazza, and R. Guzzon, "Immobilization of yeast and bacteria cells in alginate microbeads coated with silica membranes: procedures, physico-chemical features and bioactivity," Journal of Materials Chemistry, vol. 18, no. 40, pp. 4839–4848, 2008.
123. A. Navrotsky, "Technology and applications Nanomaterials in the environment, agriculture, and technology (NEAT)," Journal of Nanoparticle Research, vol. 2, pp. 321–323, 2000.
124. U. K. Parashar, P. S. Saxena, and A. Srivastava, "Role of nanomaterials in biotechnology," Digest Journal of Nanomaterials and Biostructures, vol. 3, pp. 81–87, 2008.
125. M. Auffan, J. Rose, J.-Y. Bottero, G. V. Lowry, J. P. Jolivet, and M. R. Wiesner, "Towards a definition of inorganic nanoparticles from an environmental, health and safety perspective," Nature Nanotechnology, vol. 4, no. 10, pp. 634–641, 2009.
126. K. L. Bailey, S. M. Boyetchko, and T. Längle, "Social and economic drivers shaping the future of biological control: a Canadian perspective on the factors affecting the development and use of microbial biopesticides," Biological Control, vol. 52, no. 3, pp. 221–229, 2010.
127. A. Srivastava, O. N. Srivastava, S. Talapatra, R. Vajtai, and P. M. Ajayan, "Carbon nanotube filters," Nature Materials, vol. 3, no. 9, pp. 610–614, 2004.
128. L. Chen, X. Yang, W. Raza, J. Luo, F. Zhang, and Q. Shen, "Solid-state fermentation of agro-industrial wastes to produce bioorganic fertilizer for the biocontrol of Fusarium wilt of cucumber in continuously cropped soil," Bioresource Technology, vol. 102, no. 4, pp. 3900–3910, 2011.
129. P. Wawrzyńczak, P. Białkowski, J. Rabcewicz, M. Plaskota, and B. Gotowicki, "Application of biofertilizers and biostimulants in organic orchards," in Proceedings of the Ogólnopolską Naukową Konferencję Ekologiczną „Osiągnięcia i Możliwości Rozwoju Badań i Wdrożeń w Ekologicznej Produkcji Ogrodniczej, pp. 85–86, Skierniewice, Poland, 2011.

130. Y. Bashan and H. Levanony, "Horizontal and vertical movement of Azospirillum brasilense Cd in the soil and along the rhizosphere of wheat and weeds in controlled and field environments," Journal of General Microbiology, vol. 133, pp. 3473–3480, 1987.

131. M. Albareda, M. S. Dardanelli, C. Sousa, M. Megías, F. Temprano, and D. N. Rodríguez-Navarro, "Factors affecting the attachment of rhizospheric bacteria to bean and soybean roots," FEMS Microbiology Letters, vol. 259, no. 1, pp. 67–73, 2006.

132. T. J. Avis, V. Gravel, H. Antoun, and R. J. Tweddell, "Multifaceted beneficial effects of rhizosphere microorganisms on plant health and productivity," Soil Biology and Biochemistry, vol. 40, no. 7, pp. 1733–1740, 2008.

Chapter 2

Bioconversion of Wastewater from Sweet Potato Starch Production to Paenibacillus Polymyxabiofertilizer for Tea Plants

Shengjun Xu[1], Zhihui Bai[1], Bo Jin[3], Runlin Xiao[2], and Guoqiang Zhuang[1]

[1]Research Center for Eco-Environmental Sciences, Chinese Academy of Sciences, Beijing 100085, China,

[2]Institute of Subtropical Agriculture, Chinese Academy of Sciences, Hunan 410125, China

[3]School of Chemical Engineering, The University of Adelaide, Adelaide, SA, 5005, Australia.

ABSTRACT

Wastewater from the sweet potato starch industry is a large source of nutrient-rich substrates. We assessed whether this wastewater could be used to produce *Paenibacillus polymyxa* biofertilizer for foliar application to tea trees. Using the central composite design methods we experientially determined that the optimal culture conditions for *P. polymyxa* were pH, 6.5; temperature, 29.0°C; and incubation time, 16 h. Under these conditions, a maximum biomass of 9.7×10^9 cfu/mL was achieved. We then conducted a yearlong field investigation to determine the effect of *P. polymyxa* biofertilizer on the growth of tea plants (*Camellia sinensis*). Tea yield, quantity of water extract, and tea polyphenol levels were significantly higher after foliar application of the biofertilizer compared to that in the controls by an average of 16.7%, 6.3%, and 10.4%, respectively. This approach appears to be technically feasible for organic tea production, and is an environmentally friendly way to utilize wastewater.

INTRODUCTION

Sweet potato (*Ipomoea batatas*) is one of the world's most important staple food crops and a major source of starch products used in industrial production[1]. Approximately 6 m^3 of sweet potato starch wastewater (SPSW) are produced from pre-processing of 1 ton of sweet potato roots, which is then followed by starch extraction, separation, and drying. Generation of such large volumes of SPSW has caused serious environmental problems in many Asian countries[2]. However, treatment of SPSW, which has a chemical oxygen demand (COD) of up to 25,000 mg/L, is a challenge in many developing countries.

On the other hand, SPSW could be used as an appropriate culture medium for microbial growth because of its high organic content and nontoxicity to cells. Some studies have been published on the production of microbial biomass from starch industry wastewater using various microbes such as *Aspergillus oryzae, Trichoderma viride*, and *Rhodotorula glutinis*[3, 4, 5]. To the best of our knowledge, there have been no reports on the use of SPSW for the production of *P. polymyxa*biofertilizer and the effects of foliar application of *P. polymyxa* on the growth of tea plants.

P. polymyxa (formerly *Bacillus polymyxa*) is a gram-positive, spore-forming bacterium belonging to a diverse group of plant growth-promoting bacteria (PGPB). It has a variety of beneficial effects on plants, such as enhancement of nitrogen fixation, promotion of plant growth, suppression of diseases caused by plant pathogens[6, 7, 8]. *P. polymyxa* strain EBL-06 was isolated from wheat leaves by our laboratory[9]. Since *P. polymyxa* can survive in the phyllosphere, an appropriate foliar biofertilizer can be developed, although this involves selection of a suitable substrate that supports the growth of the plant and enhances the quality of leaves. Excessive use of chemical fertilizers has increased agricultural costs, as well as causing a variety of environmental problems and concerns of food safety. For example, excessive fertilization can result in the accumulation of large amounts of nitrate in the soil, which could lead to increased leaching of nitrate into water bodies and contamination of the surface and groundwater[10, 11]. Therefore, utilization of PGPB as biofertilizers has emerged as an alternative for providing plant nutrients to increase plant yield and quality in sustainable agroecosystems[8]. Foliar application of biofertilizer avoids many of the biotic and abiotic factors and constraints of the soil environment, thereby increasing crop growth and yield. In particular, biofertilizers can meet the requirements of organic tea production, which is cultivated without the use of chemical fertilizers. In addition, the use of biofertilizers could provide an effective means of reducing chemical agricultural inputs such as synthetic pesticide, mineral fertilizers, and expensive organic fertilizers, which could lead to lower production costs.

The objectives of this study were to investigate the feasibility of using SPSW as raw material for the production of *P. polymyxa* biofertilizer and to examine whether foliar application of the biofertilizer could improve tea yield and quality. Optimal cultivation conditions for growing *P. polymyxa* were determined using the one-factor-at-a-time and central composite design (CCD) methods. We conducted a yearlong field investigation to determine if the application of *P. polymyxa*biofertilizer by foliar spray could affect the yield and quality (i.e., contents of water extract, polyphenols, caffeine, and amino acids) of tea plants (*C. sinensis*).

RESULTS

Optimization of *P. Polymyxa* Cultivation for Biofertilizer Production

The results of the one-factor-at-a-time experiments showed that the culture conditions for optimal production of *P. polymyxa* were as follows: incubation time, 16 h; pH, 7.0; temperature, 30°C; and rotational speed, 300 rpm (Fig. S1). In the CCD experiments, the optimal levels of these variables were obtained by analyzing the response surface contour plots using the software Design Expert. This facilitated the identification of the following optimal experimental conditions: pH, 6.5; temperature, 29.0°C; and rotational speed, 250 rpm (Fig. 1). The results of response surface model fitting in the form of ANOVA (analysis of variance) are shown in Table 3. Regression analysis demonstrated that the model was significant, as was evident from the calculated F-value of 17.0 and the probability value ($P = 0.0001$). Therefore, the model should be accurate for predictions within the range of the variables studied. To verify the adequacy of the model equation, three additional experiments were performed in shake flasks under the conditions derived from the model. The mean value of the obtained *P. polymyxa* biomass was 9.7×10^9 cfu/mL, which was in good agreement with the predicted value (9.8×10^9 cfu/mL). In addition, Fig. S2 shows the growth curves for the *P. polymyxa* cultivation under optimal conditions.

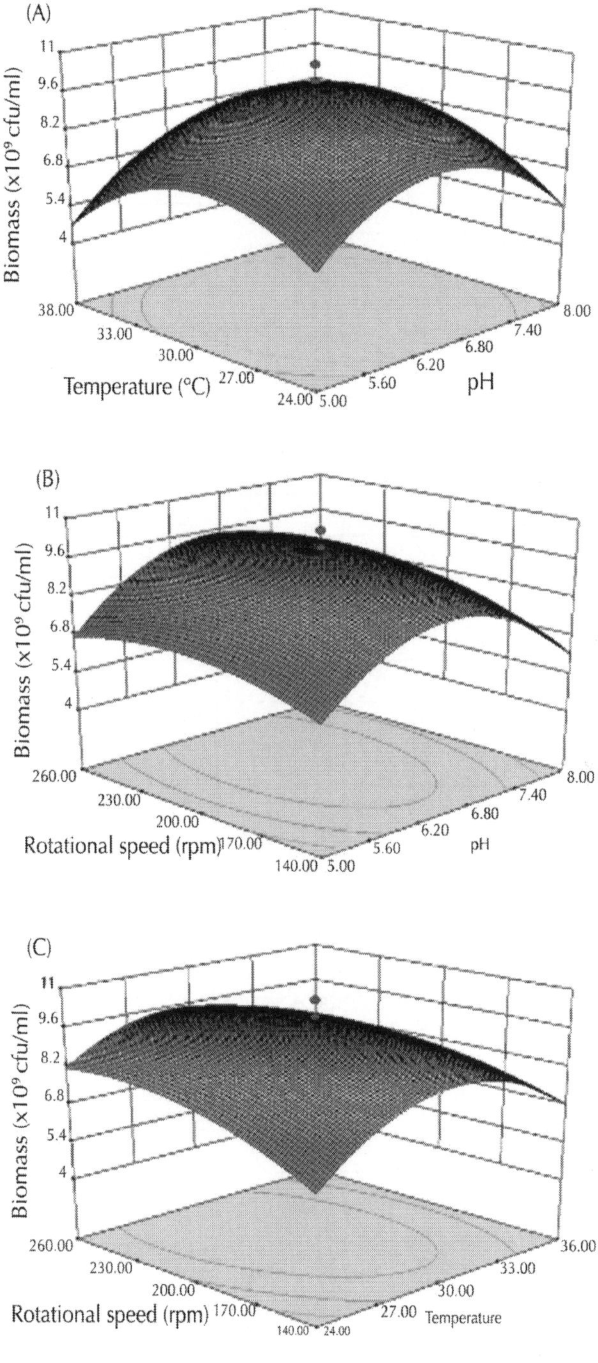

Figure 1: 3D response surface plots of the effects of mutual interactions between variables.

Table 1: Characteristics of the sweet potato starch wastewater (SPSW) and biofertilizer

Parameter (s)	SPSW	Biofertilizer	SMIA*
pH	6.3 ± 0.1	6.8 ± 0.2	5.0–8.0
Total solids (g/L)	23 ± 0.3	17 ± 0.2	-
Viable count (109/mL)	-	6.16 ± 0.36	≥0.2
Protease (U/mL)	-	37.8 ± 1.21	≥15
IAA	-	1.09 ± 0.05	-
Concentration (mg/L)			
COD	19083 ± 555	15416 ± 777	-
Total Nitrogen	1940 ± 81	1641 ± 94	-
NO3−-N	34 ± 1.7	26 ± 1.6	-
Total phosphorus	70.6 ± 4.4	67.8 ± 0.8	-
K	758 ± 58	422 ± 37	-
Fe	8.6 ± 1.1	6.5 ± 0.4	-
Mg	67.3 ± 5.5	64.0 ± 2.6	-
Zn	1.8 ± 0.06	1.6 ± 0.08	-
Cu	0.43 ± 0.09	0.11 ± 0.06	-
Cd	0.24 ± 0.02	0.23 ± 0.01	≤10
As	0.23 ± 0.05	0.21 ± 0.07	≤75
Cr	0.25 ± 0.07	0.25 ± 0.04	≤150
Pb	0.91 ± 0.05	0.87 ± 0.04	≤100

*SMIA: Standards of Microbial Inoculants in Agriculture (National Standard of China, GB 20287-2006).

Table 2: Central composite design arrangements and responses

Runs	Real levels			Biomass ×109 cfu/mL	
pH	Temperature (°C)	Rotational speed (rpm)	Actual	Predicted	
1	7 (0)	20 (−1.68)	200 (0)	4.5	4.2
2	8 (+1)	24 (−1)	140 (−1)	4.5	4.2
3	5 (−1)	24 (−1)	140 (−1)	3.2	3.9

4	5 (−1)	24 (−1)	260 (+1)	4.6	5.3
5	8 (+1)	24 (−1)	260 (+1)	6.1	6.0
6	7 (0)	30 (0)	100 (−1.68)	7.0	6.7
7	4 (−1.68)	30 (0)	200 (0)	4.0	2.6
8	7 (0)	30 (0)	200 (0)	10.0	9.8
9	7 (0)	30 (0)	200 (0)	10.6	9.8
10	7 (0)	30 (0)	200 (0)	8.9	9.8
11	7 (0)	30 (0)	200 (0)	9.7	9.8
12	7 (0)	30 (0)	200 (0)	9.6	9.8
13	9 (+1.68)	30 (0)	200 (0)	3.5	4.2
14	7 (0)	30 (0)	300 (+1.68)	9.5	9.1
15	5 (−1)	36 (+1)	140 (−1)	3.0	3.7
16	8 (+1)	36 (+1)	140 (−1)	5.0	4.9
17	5 (−1)	36 (+1)	140 (−1)	3.0	3.7
18	5 (−1)	36 (+1)	260 (+1)	4.0	4.8
19	7 (0)	40 (+1.68)	200 (0)	4.9	4.4

Table 3: Analysis of variance for the predictive equation for production of P. polymyxabiomass

Source	Sum of squares	df*	Mean Square	F-Value	Prob > F	
Model	11723.5	9	1302.6	17.0	0.0001	significant
A-pH	305.5	1	305.5	4.0	0.0768	
B-Temperature	4.4	1	4.4	0.1	0.8164	
C-Rotational speed	689.6	1	689.6	9.0	0.0149	
AB	36.1	1	36.1	0.5	0.5093	
AC	6.1	1	6.1	0.1	0.7836	
BC	3.1	1	3.1	0.0	0.8443	
A2	7059.1	1	7059.1	92.3	<0.0001	
B2	5127.8	1	5127.8	67.0	<0.0001	
C2	636.9	1	636.9	8.3	0.0180	
Residual	688.6	9	76.5			
Lack of Fit	535.4	5	107.1	2.8	0.1705	not significant

Pure Error	153.2	4	38.3			
Cor Total	12412.1	18				

*degrees of freedom.

Effect of *P. Polymyxa* Biofertilizer on the Growth of Tea Plants

P. polymyxa biofertilizer was applied to tealeaves to evaluate its effects on growth. Foliar application of the biofertilizer promoted growth in tea plants in all three seasons tested, i.e., Spring tea (SPT), Summer tea (SUT), and Autumn tea (AUT). One-hundred-bud weights of SPT, SUT, and AUT were higher than that of the control (CK) by 4.7–10.6%, 7.5–16.5%, and 7.1–21.3%, respectively (Fig. 2). We determined the optimal *P. polymyxa* concentration (4%) for producing weight increase based on the results obtained for SPT and AUT. Three repeated field experiments revealed the treatments by inoculation with the *P. polymyxa* biofertilizer at a concentration of 4% could increase the tea yield (one-hundred-bud weights) by an average of 10.5% (SPT), 21.1% (AUT), and 16.7% (SUT), respectively (Fig. 2). Compared to that of CK, the one-hundred-bud weight of SUT was significantly higher after foliar application of biofertilizers without sterilization (BO), with an average increase of 16.7%. By comparison, biofertilizers with sterilization (BS) and SPSW for SUT had less effect on plant growth; bud weights increased by 7.5% and 3.4%, respectively. Tea plant growth in response to the dose of *P. polymyxa* biofertilizer is shown in Table S1.

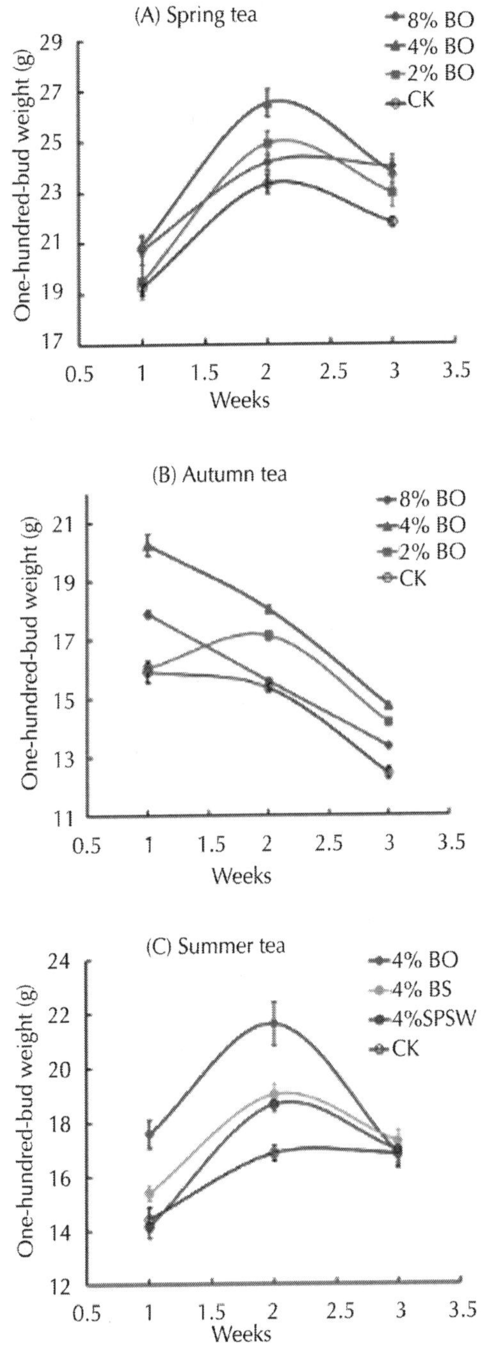

Figure 2: Effect of application of various concentrations of *P. polymyxa* biofertilizer on one-hundred-bud weight during the three seasons.

(CK, water; BO, biofertilizer; BS, sterilized biofertilizer; SPSW, sweet potato starch industry wastewater).

Effect of *P. Polymyxa* Biofertilizer on Green Tea Extracts (GTEs) and Tea Polyphenols

Levels of GTEs and tea polyphenols in tea plants treated with BS and BO were significantly higher than those in CK ($P < 0.05$). As shown in Fig. 3 and Fig. 4, the maximum GTE and tea polyphenol levels were obtained with BO, followed by BS, whereas the lowest GTE level was obtained with the SPSW and from the CK. These results were consistent with those of plant growth. Thus, foliar fertilization could be the major factor promoting tea quality. However, our results showed that concentrations of GTEs were significantly increased in SPT, SUT, and AUT by 7.4%–11.8%, 2.8%–5.2%, and 7.4%–10.6% over control (CK), respectively, while concentrations of tea polyphenols were increased by 8.6%–16.4%, 7.9%–11.3%, 18.6%–28.9% over control (CK), respectively. Overall, the data obtained from both crops showed that application of bacterial strain*P. polymyxa* as a biofertilizer for growth of *C. sinensis* could significantly enhance the physiological characters of tea compared to control. Statistical results for levels of GTEs and tea polyphenols with different treatments are shown in Table S1.

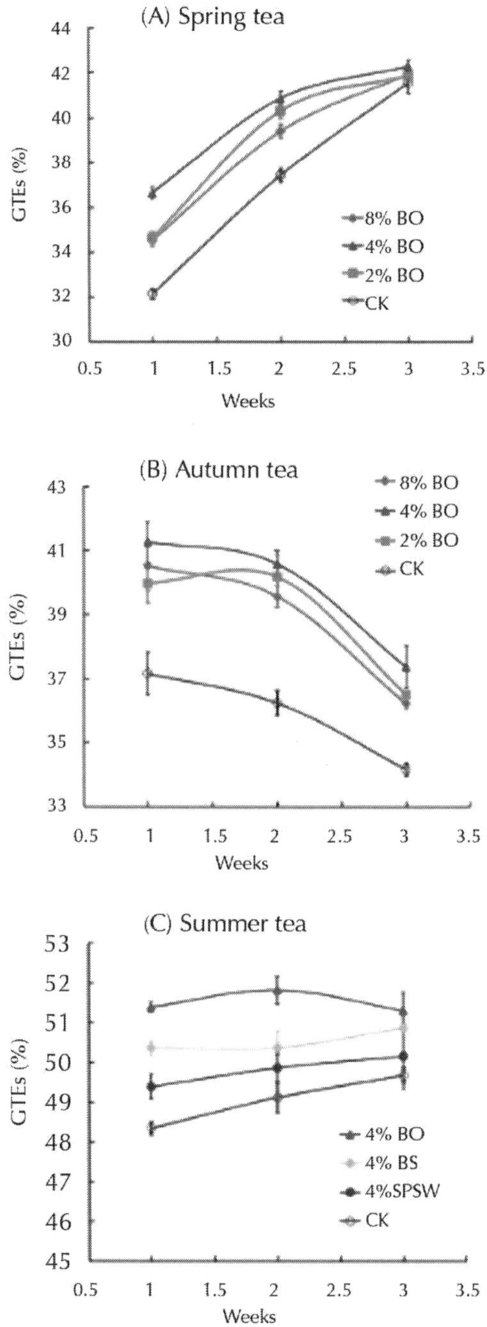

Figure 3: Effect of application of various concentrations of *P. polymyxa* biofertilizer on green tea extracts (GTEs) during three seasons.

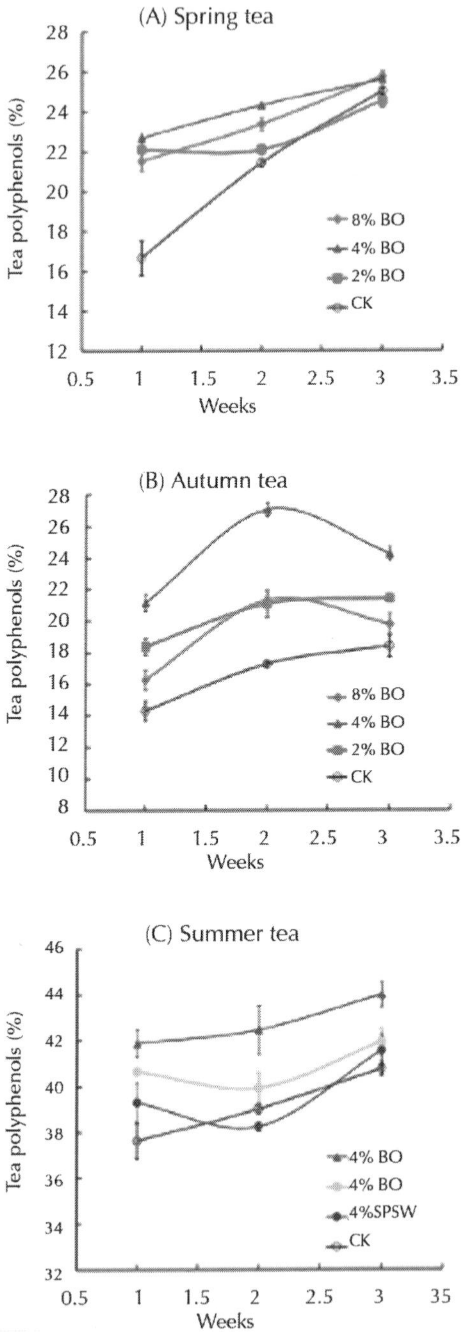

Figure 4: Effect of application of various concentrations of *P. polymyxa* biofertilizer on tea polyphenols during three seasons.

(CK, water; BO, biofertilizer; BS, sterilized biofertilizer; SPSW, sweet potato starch industry wastewater).

Effect of *P. Polymyxa* Biofertilizer on Caffeine and Amino Acids Content

Increase in concentrations of caffeine and amino acids due to *P. polymyxa* inoculation on tea in all three seasons was not statistically significant relative to that in the control (Table S1). However, regular fluctuation (increase–decrease) in amino acid content was observed in both control and treated plants throughout the study period. Mean amino acid content (4%–5%) was highest in SPT, followed by SUT and AUT, by 2–3% and 1–3%, respectively (Table S2).

DISCUSSION

SPSW has a relatively high percentage of organics (COD ≈ 19000 mg/L), total nitrogen (TN = 1940 ± 81 mg/L), total phosphorus (TP = 70.6 ± 4.4 mg/L) and other plant nutrients (K = 758 ± 58 mg/L, Fe = 8.6 ± 1.1 mg/L, Mg = 67.3 ± 5.5 mg/L and Zn = 1.8 ± 0.06 mg/L) and represents an important carbon and nutrient-rich resource. Compared to industrial wastewater, the amount of toxic compounds is considerably lower in SPSW (As = 0.23 ± 0.05 mg/L, Cd = 0.24 ± 0.02 mg/L, Cr = 0.25 ± 0.07 mg/L and Pb = 0.91 ± 0.05 mg/L). These properties make SPSW an ideal substrate for the cultivation of *P. polymyxa*. This observation is supported by a recent report showing the value of starch industry wastewater as a suitable substrate for biological conversion to more valuable products[4].

The results demonstrated that the biomass of *P. polymyxa* (9.7×10^9 cfu/mL) could be easily harvested under simple cultures during a short incubation time (16 h). The metabolites (indole-3-acetic acid [IAA], protease, and α-amylase) produced by *P. polymyxa* not only enhance the productivity of agricultural crops but also improve crop quality by increasing the content of proteins, essential amino acids, and vitamins[8]. Therefore, this approach of bioconversion of SPSW to value-added products can substantially reduce the production cost of biofertilizers, as well as lead to sustainable utilization of residues, which would be beneficial both economically and environmentally.

Foliar biofertilizers of *P. polymyxa* were investigated for possible beneficial effects on the growth of tea plants. In general, the results were in agreement with those of previously published studies that have investigated the growth response of plants to inoculation with *P. polymyxa*[12, 13]. The ability of the biofertilizer to promote tea plant growth might have three mechanisms. Firstly, the raw materials (SPSW) for the production of *P. polymyxa* biofertilizer contains much nutrient elements such as N, P, K, Fe, Mg, Zn (Table 1), which are readily absorbed by tea plants. It has been demonstrated that plant leaves are able to take up inorganic nutrients provided in aqueous forms, which are then transported to shoots and roots[14]. Based on our results, nitrogen (1900 mg/L) and phosphorus (70 mg/L) in soluble forms from the biofertilizers are readily absorbed by tea plants, thus increasing the tea growth of about 3.4%. Secondly, the increase in plant growth may be due to the supply of specific nutrients to the crop, such as the auxin metabolites (IAA) and antibiotic compounds released by *P. polymyxa*. *P. polymyxa* strains are known to produce peptide antibiotics and fusaricidins, which strongly inhibit plant pathogens such as *Aspergillus niger*, *Aspergillus flavus*, *Cladosporium fulvum*, *Candida albicans*, and *Geotrichum candidum*[15, 16]. These components may have been beneficial for stimulation of plant growth in this study which could explain the increase in plant growth by 7.5% (sterilized biofertilizer, BS). Finally, the biological activity of *P. polymyxa* could increase the yield of about 9.2%, which equal to 16.7% (biofertilizer, BO) - 7.5% (sterilized biofertilizer, BS), owing to *P. polymyxa* strains could induce systemic resistance in enhancing the disease resistance in tea plants.

The mechanisms of the *P. polymyxa* strains to promote plant growth and control a variety of pathogens that invade plants were reviewed by Raza *et al.*[17] and Lal *et al.*[18] recently. Recent investigations on the mechanisms of plant growth-promoting rhizobacteria revealed that*Pseudomonas* and *Bacillus* strains could induce systemic resistance in enhancing the disease resistance in tea plants against blister disease[19]. A similar mechanism might take place from the use of *P. polymyxa* biofertilizer for tea growth in this study. However, the induction of systemic resistance mechanism will be verified in further studies.

GTEs contain a variety of biochemical components, such as tea polyphenols, amino acids, and caffeine, all of which are closely associated with tea quality. As the leading functional component of GTEs, as well as because of their beneficial medicinal properties, tea polyphenols have gained substantial interest. *In vitro* and animal studies have suggested that antioxidants found in polyphenol-containing substances may play an important role in preventing cardiovascular disease, chronic gastritis, and some cancers[20, 21]. In this study, foliar application of biofertilizer significantly increased the concentrations of tea polyphenols, suggesting that the nutrients nitrogen and phosphorus in biofertilizers promote the synthesis of tea polyphenols. In addition, Li and Xia[22]demonstrated that an adequate amount of nitrogen and phosphorus can provide sufficient raw materials and energy for biosynthesis of tea polyphenols, increasing their content in tealeaves.

Caffeine is a purine alkaloid present in high concentrations in the tea plant[23]. The caffeine biosynthesis pathway is part of purine metabolism and is catalyzed by three *S*-adenosyl-L-methionine (SAM)-dependent *N*-methylation steps: xanthosine → 7-methylxanthosine → 7-methylxanthine → theobromine → caffeine[23]. Amino acids, which constitute 1–2% of the dry weight of the tealeaf, are synthesized from glutamic acid and ethylamine by theanine synthetase[24]. The biosynthesis of caffeine and amino acids are mainly accomplished in rhizosphere not in leaf surface which can explain less increase of caffeine and amino acids by the application of foliar biofertilizers.

In summary, production of *P. polymyxa* biofertilizer from SPSW is technically feasible and the use of this biofertilizer may be beneficial for organic tea production. Furthermore, data from field investigations revealed that foliar application of the biofertilizer can enhance yield and quality of tea production. The technical approach from this study shows the commercial potential for bioconversion of wastewater to environmental friendly, value-added biofertilizers.

METHODS

Plant Growth-Promoting Bacterium

P. polymyxa strain EBL-06 isolated from wheat leaves[9] was deposited at the China General Microbiological Culture Collection Center (CGMCC) with accession number CGMCC No. 2377. The culture was maintained on potato-dextrose agar (PDA) medium at 4°C and subcultured every 6 months.

Raw Material

Starch industry wastewater, contained approximately 20 g/L COD and 2 g/L total nitrogen, was obtained from a local sweet potato starch production company (Xiangfeng Corporation, Hunan, China). The physiochemical characteristics of SPSW are presented in Table 1.

One-Factor-at-a-Time Experiments for Biofertilizer Preparation

The starter culture and the inoculum were prepared according to our previous study[9]. *P. polymyxa*cells from PDA medium were used to inoculate 500 mL Erlenmeyer flasks containing 100 mL of sterilized (121°C, 20 min) SPSW. After 24–32 h incubation at 30°C in a rotary shaker (250 rpm), inoculum from this broth was used as a seed culture. For the optimization of biofertilizer production, relevant factors such as temperature, pH and rotational speed were investigated using the one-factor-at-a-time method. A time course experiment up to 24 h was carried out in Erlenmeyer flasks (500 mL) containing 148 mL of sterilized SPSW with 2 mL *P. polymyxa* inoculum. To determine the optimal starting pH for biofertilizer production, the pH (5–8) of the medium was adjusted by the addition of 1 M HCl and 1 M NaOH before sterilization. To determine the optimal temperature for biofertilizer production, liquid cultures were incubated at 20°C, 24°C, 28°C, 32°C, 36°C, and 40°C. For optimization of rotational speed, cultures were shaken at 50, 100, 150, 200, 250, and 300 rpm. All experiments were conducted in duplicate and key results were repeated three times to establish their validity.

CCD for Optimization of Biofertilizer Production

After determining the preliminary range of culture variables according to one-factor-at-a-time experiments, a CCD was performed with three operation parameters (X_1, temperature; X_2, initial pH; X_3, rotational speed) at five levels (–1.68, –1, 0, +1, +1.68) and 19 experiments (Table 2). Five replicates at the center point were conducted for calculating the purely experimental uncertainty variance. Data from CCD were analyzed with the following second-degree polynomial equation:

$$Y = \beta_0 + \beta_1 X_1 + \beta_2 X_2 + \beta_3 X_3 + \beta_{12} X_1 X_2 + \beta_{13} X_1 X_3 + \beta_{23} X_2 X_3 + \beta_{11} X_1^2 + \beta_{22} X_2^2 + \beta_{33} X_3^2 \quad (1)$$

where Y represents the predicted response; β_0 is an intercept; β_1, β_2 and β_3 are linear coefficients; β_{12}, β_{13} and β_{23} are cross-product coefficients; β_{11}, β_{22} and β_{33} are quadratic coefficients; and X_1, X_2 and X_3 are input variables[16].

Assays of the Physiochemical and Biochemical Properties of the Biofertilizer

Changes in pH and total solids were monitored according to standard methods[25]. The COD was determined using the close reflux and titrimetric method[25]. Total nitrogen (TN) and total phosphorus (TP) were also quantified according to methods recommended by Xi and Sun[25]. Concentrations of metals (K, Cd, Cr, Cu, Fe, Mg, As, Pb and Zn) were analyzed using a Prodigy Plasma emission spectrometer (Teledyne Leeman Labs, NH). Total bacteria were enumerated by plate count using PDA culture medium. The concentration of IAA was determined according to the procedure described by Lebuhn *et al*[26]. Protease levels were determined according to standard methods[27]. Physico-chemical characteristics of the SPSW and biofertilizer are presented in Table 1.

Plant Materials and Treatments

Field investigations were conducted in tea plantations (4 years old) of semi-tropical uplands, Hunan, China (113°19′E, 28°33′N). We selected the tea plant (*C. sinensis*) to study the efficacy of PGPB strains on tea yield and quality for three seasons (Season: mid-March to mid-October; elevation: 135 m above mean sea level; rainfall: 1386 mm; mean temperature: 16.5–34.5°C; relative humidity: 65–95%). The plantation has red soil that commonly develops on old crystalline rocks under heavy rainfall conditions. There were no obvious differences between tea plants owing to unified management practices.

The field investigations were conducted in a completely randomized design with two factors. The first factor had four levels: without biofertilizers (water, CK), SPSW only, biofertilizers with sterilization (BS), and biofertilizers without sterilization (BO). The second factor comprised the addition of biofertilizers at a rate of 0, 2, 4, or 8%. There were three replicates, each consisting of 50 bushes. The experiment was laid out in a randomized block design with a plot size of 10 m × 20 m. The experimental process was carried out in three periods: Autumn tea (mid-August to mid-October 2010, AUT), Spring tea (mid-March to mid-April 2011, SPT) and Summer tea (mid-June to mid-July 2011, SUT). In autumn and spring periods, various concentrations of BO were applied as foliar spray (30 mL/m^2), while sterile water was used as control. In the summer period, we selected the appropriate concentration (obtained from autumn and spring tea treatments) of BS or BO as foliar spray, with sterile water and SPSW serving as control.

Collection and Analysis of Tea Samples

The tea bush (bud with two leaves) was plucked at the study sites on the 1st, 2nd, and 3rd week after the application of the bio fertilizer. Tea production was measured by one-hundred-bud weight and bud density, however, in this study the tea plants were trimmed uniformly whose bud density had no obvious differences and changes during a short time. So one-hundred-bud weights can be used as indicators of biomass. As well as fresh tealeaves were processed using standard methods, including fixation, rolling, and baking to prepare the green tea by the same procedure for each sample[28].

Green Tea Water Extracts (GTEs) and Determination of Nutritional Components

Green tea powders (3 g) were extracted three times with 250 mL of boiling water at 100°C for 15 min. The infusions were filtered using a vacuum pump through perfect porosity filter paper at 55°C ± 2°C. The water extract collected were used to assess GTEs, total polyphenols, caffeine, and amino acids according to the China National Standard Methord[29, 30, 31, 32].

Statistical Analyses

Design Expert software version 7.0 (Stat-Ease Inc., Minneapolis, MN, USA) was used for the experimental designs and regression analysis of data. Data of the effect of the treatments on growth and quality of tea were evaluated by ANOVA followed by Duncan's multiple range tests using SPSS 12.0 (SPSS China, Beijing, China). Differences between treatment results were considered significant at $P < 0.05$.

ACKNOWLEDGMENTS

This work was supported by the Knowledge Innovation Program of the Chinese Academy of Sciences (Nos. KZZD-EW-09-3 and KSCX2-EW-B-1-5) and the Key Technologies R&D Program of China (No. 2012BAC25B01). We would like to thank Mr. Weidong Zhang, Dingyi Yu, Yu Zhang, and Xinghui Li for their technical support.

AUTHOR CONTRIBUTIONS

S.X. and Z.B. designed and carried out the experiments, analyzed the data, and wrote the main manuscript text; B.J. wrote the paper; R.X. and G.Z. designed the experiments. All authors reviewed the manuscript.

REFERENCES

1. Wang, R., Wang, Y., Ma, G., He, Y. & Zhao, Y. Efficiency of porous burnt-coke carrier on treatment of potato starch wastewater with an anaerobic–aerobic bioreactor. *Chem. Eng. J.*148, 35–40 (2009).
2. Chavalparit, O. & Ongwandee, M. Clean technology for the tapioca starch industry in Thailand. *J. Cleaner Prod.* 17, 105–110 (2009).

3. Jin, B., van Leeuwen, H. J., Patel, B. & Yu, Q. Utilisation of starch processing wastewater for production of microbial biomass protein and fungal α-amylase by *Aspergillus oryzae.Bioresour. Technol.* 66, 201–206 (1998).
4. Verma, M., Brar, S. K., Tyagi, R. D., Surampalli, R. Y. & Valéro, J. R. Starch industry wastewater as a substrate for antagonist, *Trichoderma viride* production. *Bioresour. Technol.*98, 2154–2162 (2007).
5. Xue, F. *et al.* Pilot-scale production of microbial lipid using starch wastewater as raw material. *Bioresour. Technol.* 101, 6092–6095 (2010).
6. Jisha, M. & Alagawadi, A. Nutrient uptake and yield of sorghum (*Sorghum bicolor L. Moench*) inoculated with phosphate solubilizing bacteria and cellulolytic fungus in a cotton stalk amended vertisol. *Microbiol. Res.* 151, 213–217 (1996).
7. Lindberg, T. & Granhall, U. Isolation and characterization of dinitrogen-fixing bacteria from the rhizosphere of temperate cereals and forage grasses. *Appl. Environ. Microbiol.* 48,683–689 (1984).
8. Ryu, C.-M., Kim, J., Choi, O., Kim, S. H. & Park, C. S. Improvement of biological control capacity of *Paenibacillus polymyxa* E681 by seed pelleting on sesame. *Biol. Control.* 39,282–289 (2006).
9. Gu, L. *et al.* Production of a newly isolated *Paenibacillus polymyxa* biocontrol agent using monosodium glutamate wastewater and potato wastewater. *J. Environ. Sci.* 22, 1407–1412(2010).
10. Chang, C.-H. &Yang, S.-S. Thermo-tolerant phosphate-solubilizing microbes for multi-functional biofertilizer preparation. *Bioresour. Technol.* 100, 1648–1658 (2009).
11. Chowdhury, M. A. H., Kouno, K., Ando, T. & Nagaoka, T. Microbial biomass, S mineralization and S uptake by African millet from soil amended with various composts. *Soil Biol. Biochem.* 32, 845–852 (2000).
12. Figueiredo, M. V., Burity, H. A., Martínez, C. R. & Chanway, C. P. Alleviation of drought stress in the common bean (*Phaseolus vulgaris L.*) by co-inoculation with *Paenibacillus polymyxa* and *Rhizobium tropici*. *Appl. Soil Ecol.* 40, 182–188 (2008).

13. Khan, Z. *et al.* Plant growth promoting rhizobacterium, *Paenibacillus polymyxa* strain GBR-1, suppresses root-knot nematode. *Bioresour. Technol.* 99, 3016–3023 (2008).
14. Fernández, V. & Eichert, T. Uptake of hydrophilic solutes through plant leaves: current state of knowledge and perspectives of foliar fertilization. *Crit. Rev. Plant Sci.* 28, 36–68 (2009).
15. Deng, Y. *et al.* Identification of LI-F type antibiotics and di-n-butyl phthalate produced by*Paenibacillus polymyxa*. *J. Microbiol. Methods* 85, 175–182 (2011).
16. Raza, W., Makeen, K., Wang, Y., Xu, Y. & Shen, Q. Optimization, purification, characterization and antioxidant activity of an extracellular polysaccharide produced by*Paenibacillus polymyxa* SQR-21. *Bioresour. Technol.* 102, 6095–6103 (2011).
17. Raza, W., Yang, W. & Shen, Q.-R. *Paenibacillus polymyxa*: antibiotics, hydrolytic enzymes and hazard assessment. *J. Plant Pathol.* 93, 419–430 (2008).
18. Lal, S. & Tabacchioni, S. Ecology and biotechnological potential of *Paenibacillus polymyxa*: a minireview. *Indian J. Microbiol.* 49, 2–10 (2009).
19. Saravanakumar, D., Vijayakumar, C., Kumar, N. & Samiyappan, R. PGPR-induced defense responses in the tea plant against blister blight disease. *Crop Prot.* 26, 556–565 (2007).
20. Chen, Q., Zhao, J., Liu, M., Cai, J. & Liu, J. Determination of total polyphenols content in green tea using FT-NIR spectroscopy and different PLS algorithms. *J. Pharm. Biomed. Anal.*46, 568–573 (2008).
21. Khan, N. & Mukhtar, H. Tea polyphenols for health promotion. *Life sciences* 81, 519–533(2007).
22. Li, j. & Xia, J. Summary on Nitrogen (N), Phosphorus (P), Potassium (K) and Tea Quality.*Chinese Agricultural Science Bulletin* 21, 62 (2005).
23. Shi, C. *et al.* Deep sequencing of the *Camellia sinensis* transcriptome revealed candidate genes for major metabolic pathways of tea-specific compounds. *BMC genomics* 12, 131(2011).
24. Sasaoka, K., Kito, M. & Inagaki, H. Studies on the biosynthesis of theanine in tea seedlings: synthesis of theanine by the homogenate of tea seedlings. *Agric. Biol. Chem.* 27, 467–468(1963).

25. Xi, D. & Sun Y. . *Environmental Monitoring,* [94–96, 70–71, 113–118, 104–110, 112–113] (Higher Education Press, Beijing, 2010).

26. Lebuhn, M., Heulin, T. & Hartmann, A. Production of auxin and other indolic and phenolic compounds by *Paenibacillus polymyxa* strains isolated from different proximity to plant roots. *FEMS Microbiol. Ecol.* 22, 325–334 (1997).

27. General Administration of Quality Supervision, Inspection and Quarantine of P. R. China. & Standardization Administration of P. R. China. *Microbial inoculants in agriculture.* (China Standards Press, Beijing, 2006).

28. General Administration of Quality Supervision, Inspection and Quarantine of P. R. China. & Standardization Administration of P. R. China. . *Production technical practice for tea.* (China Standards Press, Beijing, 2011).

29. Zhou, W. *et al.* [Tea — Determination of water extracts content] *Standards collection for tea.*[The First Editing Group of China Standards Press (ed.)] [502–503] (China Standards Press, Beijing, 2011).

30. Zhou, W. *et al.* [Determination of total polyphenols and catechins content in tea] *Standards collection for tea.* [The First Editing Group of China Standards Press (ed.)] [533–539] (China Standards Press, Beijing, 2011).

31. Zhou, W. *et al.* [Tea — Determination of caffeine content] *Standards collection for tea.* [The First Editing Group of China Standards Press (ed.)] [529–531] (China Standards Press,2011).

32. Zhou, W. *et al.* [Tea — Determination of free amino acids content] *Standards collection for tea.* [The First Editing Group of China Standards Press (ed.)] [541–542] (China Standards Press, Beijing, 2011).

Chapter 3

Optimization of Dairy Sludge for Growth of Rhizobium Cells

Ashok Kumar Singh, Gauri Singh, Digvijay Gautam, and Manjinder Kaur Bedi

Department of Microbiology, Dolphin (PG) Institute of Biomedical and Natural Sciences, Manduwala, Dehradun, Uttarakhand, India

ABSTRACT

In this study dairy sludge was evaluated as an alternative cultivation medium for Rhizobium. Growth of bacterial strains at different concentrations of Dairy sludge was monitored. Maximum growth of all strains was observed at 60% Dairy sludge concentration. At 60% optical density (OD) values are 0.804 for Rhizobium trifolii (MTCC905), 0.825 for Rhizobium trifolii (MTCC906), and 0.793 for Rhizobium meliloti (MTCC100). Growth pattern of strains was observed at 60% Dairy sludge along with different synthetic media (tryptone yeast,Rhizobium minimal medium and yeast extract mannitol). Growth in 60% Dairy sludge was found to be superior to standard media used for Rhizobium. Media were optimized using 60% dairy sludge along with different concentrations of yeast extract (1–7 g/L) and mannitol (7–13 g/L) in terms of optical density at different time intervals, that is, 24, 48 and 72 hours. Maximum growth was observed in 6 g/L of yeast extract and 12 g/L of mannitol at 48-hour incubation period in all strains. The important environmental parameters such as pH were optimized using 60% dairy sludge, 60% dairy sludge +6 g/L yeast extract, and 60% dairy sludge +12 g/L mannitol. The maximum growth of all strains was found at pH 7.0. The present study recommends the use of 60% dairy sludge as a suitable growth medum for inoculant production.

INTRODUCTION

The first step in the production of legume inoculants is massive growth of a selected Rhizobial strain in liquid medium [1]. As in case of many other industrial fermentations, the economy of such a process is largely governed by the price of media utilized. The suitability of culture media for large-scale production of Rhizobium depends upon the utilization of carbon source and multiplication rate of bacteria. For industrial production of rhizobial inoculants, it is important to identify inexpensive and easily available sources of nutrients for culture medium. Nutrient media such as yeast extract mannitol, tryptone yeast extract, and rhizobial minimal media are found to be very suitable for the growth of rhizobia. The standard medium includes mannitol, sucrose or glycerol as the carbon source, yeast extract as a source of nitrogen, growth factors, and mineral salts. The YMB medium (Yeast Mannitol Broth) has been mostly used for a laboratory-scale production [2–4]; however, its industrial use is limited due to high cost [5].

In view of the growing demand of rhizobial inoculants it is important to search cheap and readily available substances against these expensive ingredients. Many investigators have looked for ways of producing biofertilizers using low-cost media. A variety of agricultural and industrial byproducts such as proteolyzed pea husks and water hyacinth [6], malt sprouts [7], deproteinized leave extracts [8], bagasse [9], waste water sludge [10], Jaggery solution [11], and sugar waste [12] have been used for commercial production of inoculants. All these materials contain growth factors, nitrogen and carbon for growth of various strains of Rhizobium equal to or better than the known growth in the available media.

In the present study, this aim was achieved for fast-growing rhizobia, using the dairy industry sludge as a suitable growth medium. Rhizobium meliloti (MTCC100), Rhizobium trifolii(MTCC 905), and Rhizobium trifolii (MTCC 906) strains obtained from Microbial Type Culture Collection (MTCC) from the Institute of Microbial Technology (IMTECH), Chandigarh, India were used as model strains.

The effluents are generated from milk processing through milk spillage, drippings, washing of cans, tankers bottles, utensil, equipments, and floors. The dairy industry generates an average of 2.5–3.0 liters of wastewater per liter of milk processed. Dairy sludge contains casein, lactose, fat, and valuable N, P, K, and organic matter. Dairy sludge has lower levels of heavy metals and other harmful components than waste water sludge and molasses used for cultivatingRhizobium.

MATERIALS AND METHODS

Strains and Substrate used

Rhizobium meliloti and Rhizobium trifolii strains were used. The dairy sludge was collected from Anchal Dairy which is situated in Dehradun on Sehestra Dhara Road. It is an effluent from dairy plant that is not treated and released as such. This dairy is well known in Dehradun for milk products like curd, cheese, and butter.

Starter Culture

50 mL of culture was prepared by inoculating Rhizobium strains into Yeast Extract Mannitol Broth (YEM). Starter culture was sterilized through autoclaving and incubated at 28 ± 2°C for 24 hrs.

Optimization of Dairy Sludge and Comparison with Different Lab Media

1% of starter culture was inoculated into different concentrations of Dairy sludge, that is, 2%, 4%, 6%, 8%, 10%, and 20% up to 100% and growth was monitored by recording optical density (OD) at regular time intervals.

Effect of different media that is, 60% Dairy sludge, RMM, TY, and YEM on Rhizobium was monitored in terms of absorbance at 600 nm after every 24 hrs up to 72 hrs by inoculating the bacteria with these media.

Optimization of Dairy Sludge with Components of Standard Media

Yeast extract and mannitol are two main components of the standard media which are very helpful for the growth of Rhizobium. So the growth of bacteria was also observed in 60% sludge along with different concentrations of yeast extract (1 gm/L, 2 gm/L, and 3 gm/L up to7 gm/L) and similarly in 60% sludge along with different concentrations of mannitol (7 gm/L, 8 gm/L, and 9 gm/L up to 12 gm/L).

Effect of PH on Growth

Rhizobium strains were inoculated into 60% dairy sludge, 60% dairy sludge +6 g/L yeast extract and 60% dairy sludge +12 g/L mannitol as suitable media, maintained at different pH, that is, 6.0, 6.5, 7.0, 7.5, and 8.0, and growth was monitored in terms of absorbance at 600 nm after every 24 hrs up to 72 hrs.

RESULTS

Effects of Concentration of Dairy Sludge on Growth of Rhizobium Strains

Growth of Rhizobium trifolii (MTCC905), Rhizobium trifolii (MTCC906), and Rhizobium meliloti (MTCC100) at different concentrations of Dairy sludge (2%, 4%, 6%, 8%, 10%, 20%, 30%, 40%, 50%, 60%, 70%, 80%, 90%, and 100%) was monitored by recording optical density (OD) at 600 nm after 48 hrs incubation. There was a considerable decline in OD values after 60% Dairy sludge concentration. At 60% concentration OD values are 0.804 for Rhizobium trifolii (MTCC905), 0.825 for Rhizobium trifolii (MTCC906), and 0.793 for Rhizobium meliloti(MTCC100). However effect of Dairy sludge concentration (above 60%) was clearly visible on the growth of above three strains (Figure 1). Minimum growth for Rhizobium trifolii(MTCC905) and Rhizobium trifolii (MTCC906) was observed at 2.0% Dairy sludge concentration that is, 0.071 and 0.047, respectively, while minimum growth of Rhizobium meliloti (MTCC100) was observed at 10% Dairy sludge concentration, that is, 0.133.

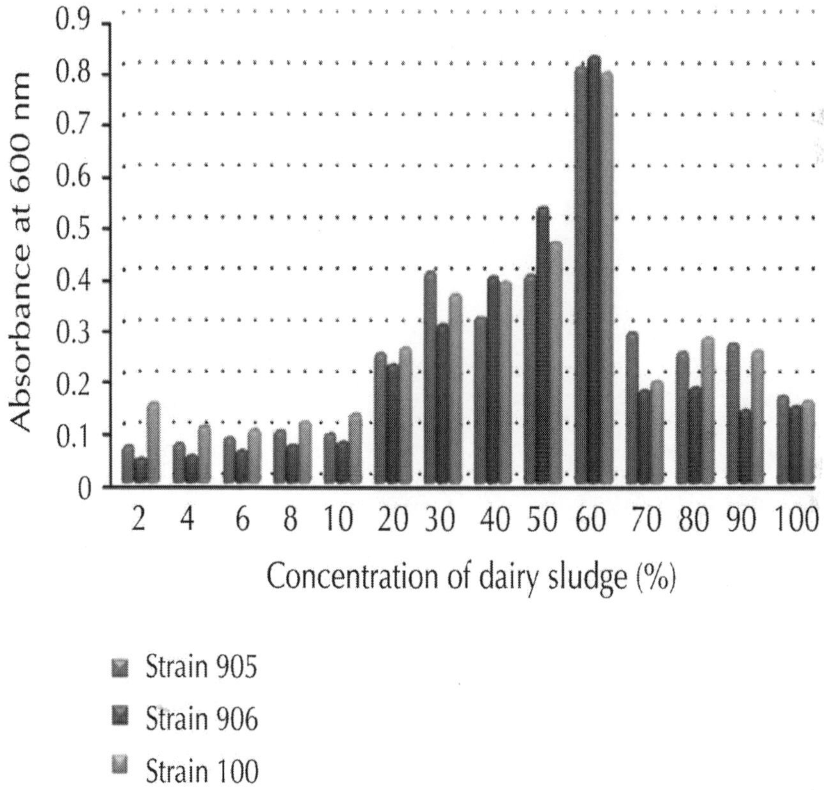

Figure 1: Growth of Rhizobium on different concentrations of dairy sludge.

Growth of Rhizobium Strainson Dairy Sludge Concentration and Synthetic Medium (YEM, RMM, and TY Media)

Growth of Rhizobium strains was observed at 60% Dairy sludgeconcentration along with YEM, RMM, and TY Media separately within the same conditions. Growth of Rhizobium strains was found maximum in 60% Dairy sludge concentration at 48 hrs, that is, 0.625 in Rhizobium trifolii (MTCC905), 0.804 in Rhizobium trifolii (MTCC906), and 0.793 in Rhizobium meliloti(MTCC100) and found minimum in RMM Medium in all strains used in the current study (Figure 2).

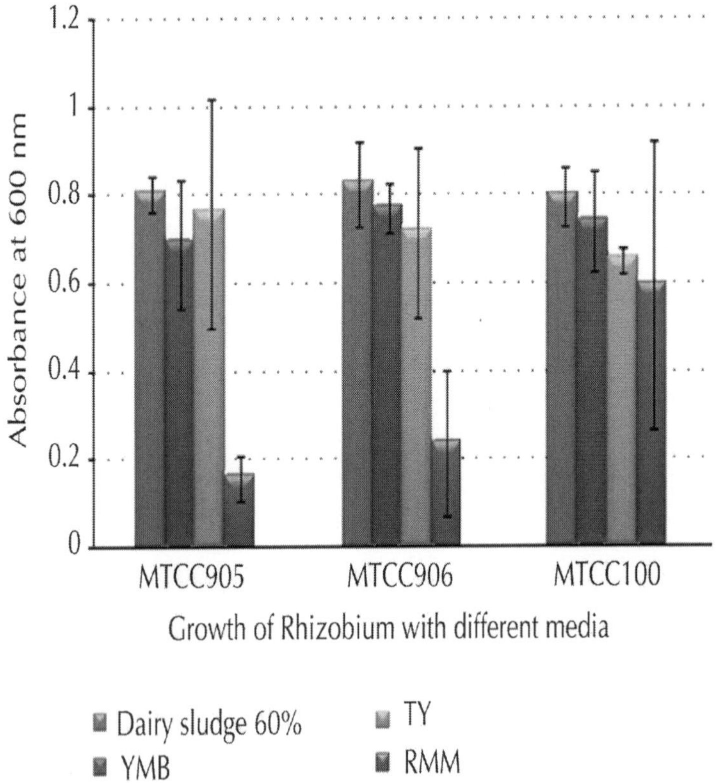

Figure 2: Comparison of growth of Rhizobium on different media with 60% dairy sludge.

Media Optimization Using 60% Dairy Sludge along with Different Concentrations of Yeast Extract and Mannitol

Rhizobium strains were exposed to various concentrations of yeast extract (1–7 g/L) and mannitol (7–13 g/L) in terms of optical density at different time intervals, that is, 24, 48, and 72 hours. Rhizobial cell viability was maintained throughout the 72-hour growth period. Maximum growth was observed in 6 g/L of yeast extract (Figure 3) and 12 g/L of mannitol (Figure 4) at 48-hour incubation period in all strains.

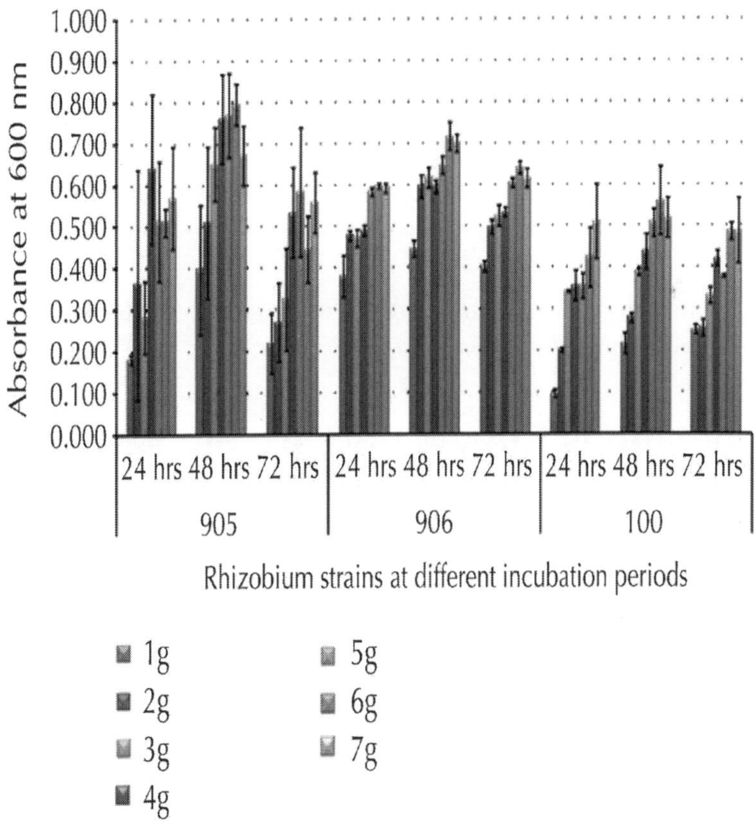

Figure 3: Growth of Rhizobium on 60% dairy sludge at various concentrations of yeast extract at different incubation periods.

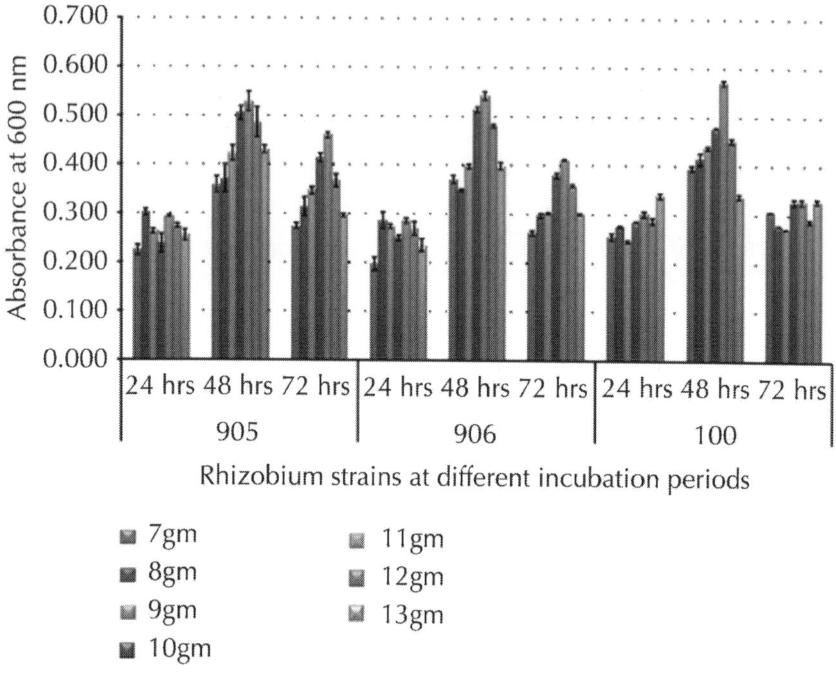

Figure 4: Growth of Rhizobium on 60% dairy sludge at various concentrations of mannitol at different incubation periods.

Combinations Effects of Dairy Sludge, Yeast Extract, and Mannitol at Different PH

At 28°C and 60% of dairy sludge, a considerable increase in OD values was observed reaching a maximum of 0.808 ± 0.001 in Rhizobium trifolii (MTCC905), 0.795 ± 0.006 in Rhizobium trifolii (MTCC906), and d 0.762 ± 0.004 in Rhizobium meliloti (MTCC100) with increasing pH up to 7.0 in 48 hrs (Figure 5).

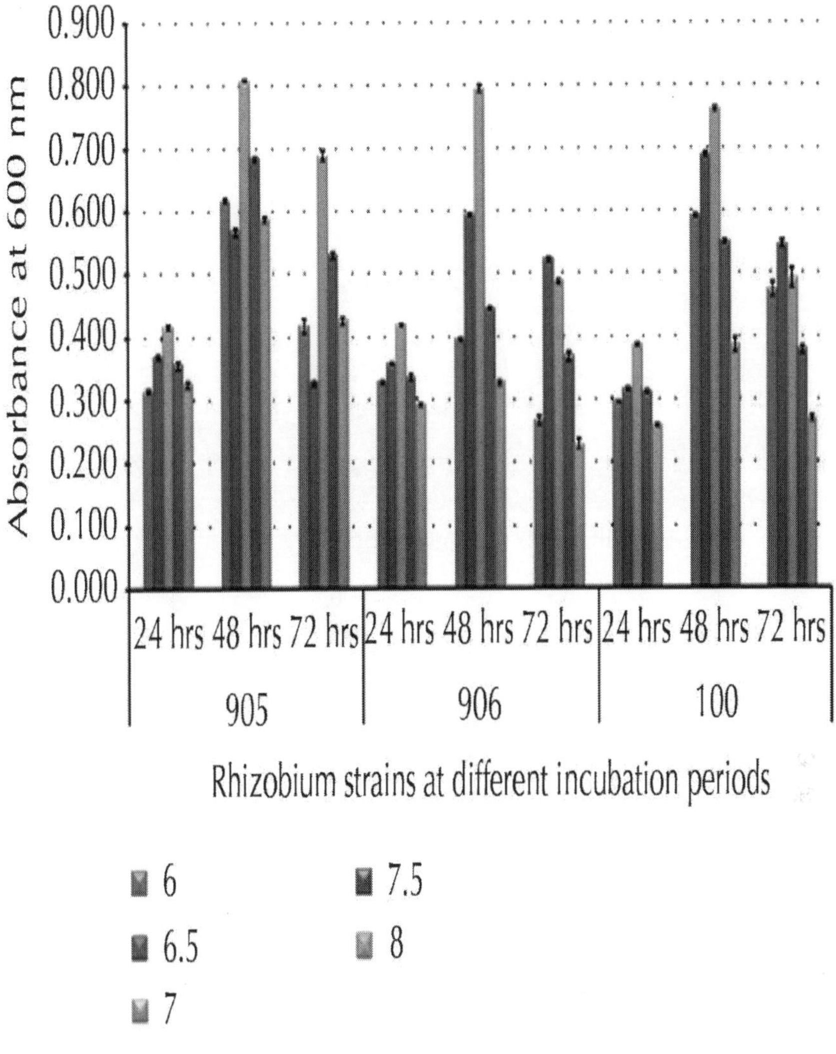

Figure 5: Growth of Rhizobium on 60% dairy sludge at various substrate at different incubation period.

At 28°C and 6.0 g/L yeast extract, growth of Rhizobium strains at different pH was recorded in terms of optical density (600 nm) at different time intervals (24, 48, and 72 hours) (Figure 6). There was a significant increase in the OD values reaching a maximum of 0.689 ± 0.003 in Rhizobium trifolii (MTCC905), 0.693 ± 0.023 in Rhizobium trifolii (MTCC906), 0.488 ± 0.003 and in Rhizobium meliloti (MTCC100) with the increase in pH that is up to 7.0 in 48 h.

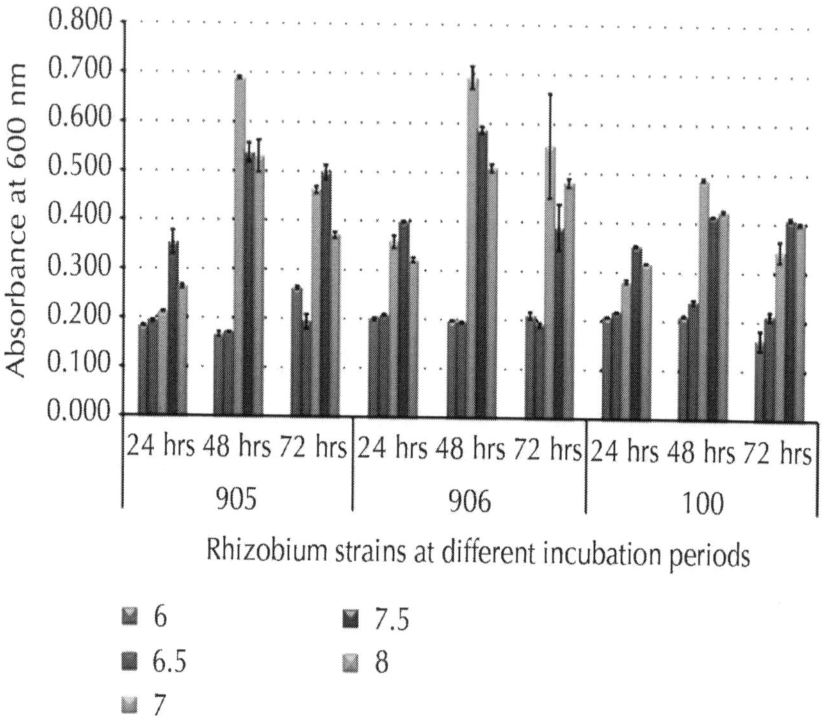

Figure 6: Growth of Rhizobium on 60% dairy sludge with yeast extract with different pH at different incubation periods.

At 28°C and concentration of mannitol 12 g/L. growth of Rhizobium trifolii (MTCC905), Rhizobium trifolii (MTCC906), and Rhizobium meliloti (MTCC100) at different pH was recorded. There was an increase in the OD values reaching a maximum of 0.658 ± 0.003 inRhizobium trifolii (MTCC905), 0.683 ± 0.010 in Rhizobium trifolii (MTCC906), and 0.612 ± 0.010 in Rhizobium meliloti (MTCC100) with the increase in pH that is up to 7.0 in 48 h. However, effect of pH (below and above 7.0) was clearly visible on the growth of strains (Figure 7).

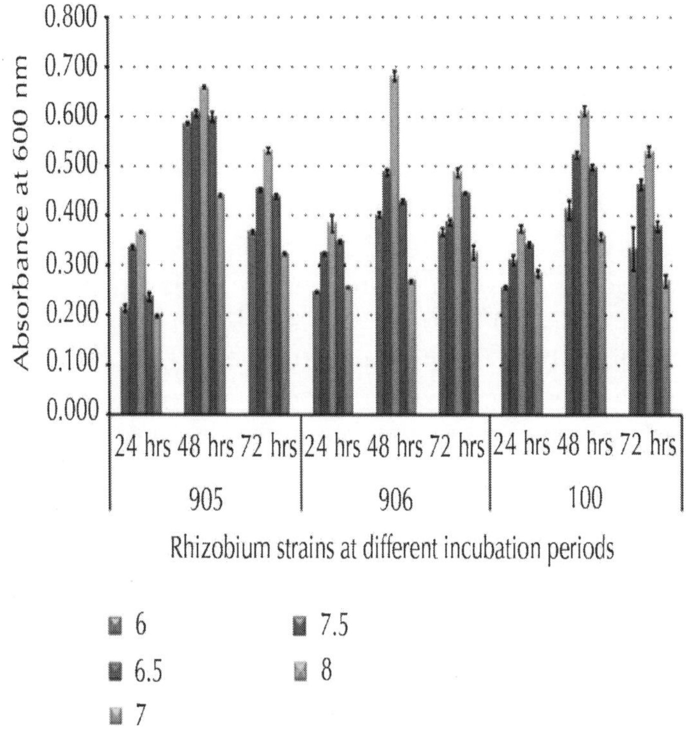

Figure 7: Growth of Rhizobium on 60% dairy sludge with mannitol at various pH with different incubation periods.

DISCUSSION

Dairy sludge obtained from Anchal Dairy which is situated in Dehradun (U.K.) is used as a growth media for fast growing rhizobia. Growth of R. trifolii and R. meliloti was observed at different concentrations of dairy sludge (2%, 4%, 6%, 8%, 10%, and 20% up to 100%) by recording optical density (OD) at 600 nm. Growth of R. trifolii and R. meliloti was maximum at 60% concentration. Similar work has been done using other substrates. Singh et al. [13] optimize different concentrations of sugar waste for fast-growing rhizobia and found maximum growth at 10% sugar waste. Ben Rebah et al. [10] demonstrated waste water sludge as a culture medium for rhizobia.

In the above study growth of all strains was increasing up to 60% dairy sludge, and above 60% concentration, fall in growth of Rhizobium was observed, which might be due to the fact that at higher concentrations of dairy sludge aeration conditions are improper in the medium. The aerobic condition is good for nitrogen-fixing bacteria (Hassen et al.) [14]. Therefore 60% concentration of dairy sludge was taken as optimum for further study.

The results of this study indicates that on medium containing only 60% dairy sludge, cells of fast-growing rhizobia (R. meliloti and R. trifolii) grow rapidly and growth was superior to that of the control (standard media). These results demonstrate that 60% dairy sludge can be satisfactorily used as a growth substrate of rhizobia. Annapurna et al. [15] used 1% aqueous solutions of commercial quality molasses, malt extract, jaggery, peptone, and yeast extract as the sole source of nutrients for the two species of Rhizobium, that is, R. trifolii (RC 1–4), a fast grower, and R. Japonicum (SB-16), a slow grower. YEM Broth was used as standard medium for comparison; among different media used jaggery solution supported maximum growth (10.17 for R. trifolii and 9.98 for R. Japonicum) followed by molasses (9.65 for R. trifolii and 9.22 for R. Japonicum) though it was not comparable to that of YEM Broth which supported 10.31 for R. trifolii and 10.13 for R. Japonicum; therefore, according to this finding jaggery solution and molasses can be used as an alternate of YEM Broth in commercial preparation of inoculants. This will considerably reduce the production cost. Similar observation has been reported by Daniel et al. (2009) [16]. The composition of the alternative culture media included glycerol, molasses, glutamate, yeast extract and salts for Rhizobium was optimized and no significant difference was observed in the alternate culture media, the traditional one (Yeast extract mannitol agar medium). Estrella et al. [17] found that cheese whey-based growth medium efficiently protects Rhizobium loti cells during freezing and can maintain their growth in a manner similar to that of traditional mannitol-based medium (YEM).

The concept of media optimization was to establish a medium which shows the optimum conditions for the growth of the organism at cheap cost as compared to normal media.Rhizobium strains were able to utilize glucose and sucrose more efficiently than normal YEM medium [18]. Yeast extract and mannitol are two main components of the standard media for the growth of Rhizobium; therefore, 60% sludge along with different concentrations of yeast extract (1 gm/L, 2 gm/L, and 3 gm/L up to 7 gm/L) and 60% sludge with different concentrations of mannitol (7 gm/L, 8 gm/L, and 9 gm/L up to 13 gm/L) were further optimized. Growth in both conditions was studied at different time intervals that is, 24, 48, and 72 hours. Our results demonstrated that maximum growth was observed in 6 g/L of yeast extract and 12 g/L of mannitol at 48-hour incubation period in all strains.

The effects of additional sources of nutrients on the growth of S. meliloti were studied in secondary sludge from the Quebec City Wastewater Plant (Ben Rebah et al.) [19]. Sludge supplemented with different concentrations of yeast extract and glycerol increased the maximum cell counts. The highest increase was approximately sixfold and was obtained with the addition of 4 g/L yeast extract and 7.5 g/L glycerol.

pH is an important parameter for the growth of organisms, slight variation in pH of medium might have enormous effects on the growth of organisms, and keeping this in mind, growth of fast growing strains was monitored at different pH (6.0, 6.5, 7.0, 7.5, and 8.0) at different time intervals, that is, 24, 48, and 72 hours. Conditions used were 60% dairy sludge, 60% dairy sludge +6 g/L yeast extract, and 60% dairy sludge +12 g/L mannitol.

The result of the effects of different pH values on the growth of rhizobia shows that these strains grow well at pH 7.0 on observing OD at 48 hrs incubation period. This is in accordance with Mensah et al. [20]; they recorded maximum absorbance for broth experiment as well as exhibited heavy growth as measured by population count at pH 7.0 for Rhizobium species. Ali et al. [21] also observed that there was a considerable increase in OD values with increasing pH up to 7.0. Rodrigues et al. [22] reported that pH 7.0 is the most optimum pH for the growth of root modulating bacteria. The above results indicate that optimum growth occurred at pH 7.0 and incubation period of 48 Hrs.

CONCLUSIONS

Our study demonstrated for the first time that sludge generated by dairy industry sustains growth of fast-growing rhizobia higher than standard medium and different components used in YEM medium. The present study concludes that 60% dairy sludge is a suitable growth medium.

Production cost of biofertilizers will be reduced by reusing the dairy industry sludge as a substrate.

REFERENCES

1. J. A. Thompson, "Australian quality control and standards," in Report of the Expert Consultation on Legume Inoculant Production and Quality Control, J. A. Thompson, Ed., pp. 107–111, Food and Agriculture Organization of the United Nations, Rome, Italy, 1991.
2. J. P. Verma, J. Yadav, and K. N. Tiwari, "Application of Rhizobium sp. BHURC01 and plant growth promoting Rhizobactria on nodulation, plant biomass and yields of chickpea (Cicer arietinum L.)," International Journal of Agricultural Research, vol. 5, no. 3, pp. 148–156, 2010.
3. A. I. Fahmi, H. H. Nagaty, R. A. Eissa, and M. M. Hassan, "Effects of salt stress on some nitrogen fixation parameters in Faba Bean," Pakistan Journal of Biological Sciences, vol. 14, no. 6, pp. 385–391, 2011.
4. J. Yadav, J. P. Verma, V. K. Rajak, and K. N. Tiwari, "Selection of effective indigenousRhizobium strain for seed inoculationof chickpea (Cicer aritenium L.) production,"Journal of Bacteriology, vol. 1, pp. 24–30, 2011.
5. F. Ben Rebah, R. D. Tyagi, and D. Prévost, "Wastewater sludge as a substrate for growth and carrier for rhizobia: the effect of storage conditions on survival of SinoRhizobiummeliloti," Bioresource Technology, vol. 83, no. 2, pp. 145–151, 2002.
6. S. L. Gulati, "New nonsynthetic medium for Rhizobium culture production from wastes," Biotech and Bioengine, vol. 21, pp. 1507–1515, 1979.
7. S. V. Chanda, A. K. Joshi, P. P. Vishnav, and Y. D. Singh, "Growth analysis using classical and curve-fitting methods in relation to productivity in pearl millet (Pennisetum americanum L.)," Journal of Agronomy and Crop Science, vol. 159, pp. 312–319, 1987.
8. J. L. Bioardi and R. J. Ertola, "Rhizobium biomass production in batch and continuous culture with a malt-sprouts medium,"

Journal of Applied Microbiology and Biotechnology, vol. 1, no. 2, pp. 163–172, 1985.

9. F. R. Malik, S. Ahmed, and M. Y. Rizki, "Utilization of lignocellulosic waste for the preparation of nitrogenous Biofertilizer," Pakistan Journal of Biological Sciences, vol. 10, pp. 1217–1220, 2001.
10. F. Ben Rebah, D. Prévost, A. Yezza, and R. D. Tyagi, "Agro-industrial waste materials and wastewater sludge for rhizobial inoculant production: a review," Bioresource Technology, vol. 98, no. 18, pp. 3535–3546, 2007.
11. S. K. Jain, D. V. Pathak, and H. R. Sharma, "Alternate carbon substrate for mass production of Rhizobium inoculants," Haryana Agricultural University Journal of Research, vol. 30, pp. 1–6, 2000.
12. A. K. Singh, G. Singh, R. P. Bhatt, S. Pant, A. Naglot, and L. Singh, "Sugars waste, an alternative growth and complete medium for fast growing Rhizobium cells," African Journal of Microbiology Research, vol. 5, no. 20, pp. 3289–3295, 2011.
13. A. K. Singh, G. Singh, R. P. Bhatt, and S. Pant, "Optimization and comparative study of the sugar waste for the growth of Rhizobium cells along with traditional laboratory media," Research Journal of Microbiology, vol. 6, no. 9, pp. 715–723, 2011.
14. A. A. Hassen, J . Xu, and J . Yang, "Growth conditions of associative nitrogen-fixing bacteria Enterobacter cloace in rice plants," Agricultural Journal, vol. 2, no. 6, pp. 672–675, 2007.
15. K. Annapurna, G. Mala, and K. S. Jauhri, "Jaggery solution—acomplete nutrient medium for multiplication of Rhizobium," Current Science, vol. 57, pp. 1188–1189, 1988.
16. T. Daniel Fernando Rojas, R. Maria Fernand Garri do, and B. Ruth Rebeca Bonilla, "Standarization of a complex culture media for multiplication of C 50 Rhizobium sp. Strain," Revista Corpoica—Ciencia Y Technologia Agropecuria, vol. 10, no. 1, pp. 70–80, 2009.
17. M. J. Estrella, F. L. Pieckenstain, M. Marina, L. E. Díaz, and O. A. Ruiz, "Cheese whey: an alternative growth and protective medium for Rhizobium loti cells," Journal of Industrial Microbiology and Biotechnology, vol. 31, no. 3, pp. 122–126, 2004.

18. Ç. Küçük, M. Kivanç, and E. Kinaci, "Characterization of Rhizobium Sp. isolated from bean," Turkish Journal of Biology, vol. 30, no. 3, pp. 127–132, 2006.
19. F. Ben Rebah, R. D. Tyagi, and D. Prévost, "Production of S. meliloti using wastewater sludge as a raw material: effect of nutrient addition and pH control," Environmental Technology, vol. 23, no. 6, pp. 623–629, 2002.
20. J. K. Mensah, F. Esumeh, M. Iyamu, and C. Omoifoc, "Effects of different salt concentrations and pH on growth of Rhizobium sp. And a cowpea-rhizobium association," American-Eurasian Journal of Agricultural & Environmental Sciences, vol. 3, pp. 198–202, 2006.
21. S. F. Ali, L. S. Rawat, M. K. Meghvansi, and S. K. Mahna, "Selection of stress—tolerant rhizobial isolates of wild legumes growing in dry regions of rajasthan, India," ARPN Journal of Agricultural and Biological Science, vol. 4, pp. 13–18, 2009.
22. C. S. Rodrigues, M. Laranjo, and S. Oliverira, "Effect of heat and pH stress in the growth of chick pea mesorhizobia," Current Microbiology, vol. 53, no. 1, pp. 1–7, 2006.

Chapter 4

Vermiculture Technology: Reviving the Dreams of Sir Charles Darwin for Scientific Use of Earthworms in Sustainable Development Programs

Rajiv K. Sinha[1*], Sunita Agarwal[2], Krunal Chauhan[3], Vinod Chandran[3], Brijal Kiranbhai Soni[3]

[1*]Griffith School of Engineering (Environment), Griffith University, Nathan Campus, Brisbane, Australia

[2]University of Rajasthan, Jaipur, India

[3]Vermiculture Project, Griffith University, Brisbane, Australia

ABSTRACT

Vermiculture technology is emerging as an "environmentally sustainable", "economically viable" and "socially acceptable" technology all over the world. 1) Vermi-composting Technology (to manage most organic wastes); 2) Vermi-filtration Technology (to treat municipal & several industrial wastewater); 3) Vermiremediation Technology (to treat & clean up contaminated lands); 4) Vermi-agro-production Technology (to produce chemical-free organic foods by worms & vermicompost); 5) Vermi-industrial Production Technology (to produce valuable industrial raw materials from worms). The use of earthworms as "waste managers" for efficient "composting of food and farm wastes" and as "soil managers" for "fertility improvement" and enhanced "farm production" were known for ages but now it is being more scientifically and also commercially revived. The other uses of earthworms for the benefits of environment and society (wastewater treatment, land remediation & production of valuable medicines even to combat cancer and heart diseases; raw materials for rubber, lubricant, soap, detergent & cosmetic, industries and protein rich feed materials for fishery, dairy & poultry industries are some "new discoveries". We have successfully experimented with the first four technologies for management of "municipal solid wastes", treatment of "municipal & industrial wastewater", remediation of "PAHs contaminated soils" and production of "wheat & corn crops" by use of vermicompost at Griffith University, Australia, with excellent results. Wastes are degraded by over 75% faster than conventional systems and compost produced are disinfected, detoxified, richer in nutrients & beneficial soil microbes; BOD loads & TSS of wastewater is reduced by over 95%; PAHs from contaminated soils are removed by over 80% in just 12 weeks; and crops growths are promoted by 30-40% higher as compared to chemical fertilizers. Earthworms are both "protective" & "productive" for environment and society.

INTRODUCTION

The global scientific community today is searching for a technology which should be "economically viable" (cheaper to be afforded by all nations), "environmentally sustainable" (friendly to the environment—

flora, fauna, soil, air & water, with no adverse effect on them) and "socially acceptable" (beneficial to the society with no adverse effect on human health). Vermiculture Technology combines all these virtues and qualities together.

A revolution is unfolding in vermiculture studies (rearing of useful earthworms species) for multiple uses in environmental protection and sustainable development [1-3]. Earthworms have over 600 million years of experience as "ecosystem engineers". Vermiculture scientists all over the world knew about the role of earthworms as "waste managers", as "soil managers & fertility improvers" and "plant growth promoters" for long time. But some comparatively "new discoveries" about their role in "wastewater treatment", "contaminated soil remediation", and more recently about their potential use in modern medicine for protection of "human health" such as in lowering of blood pressure, thinning of blood and dissolving blood clots for stroke and heart patients, cure for cancer, cure for arthritis & rheumatisms, as an anti-inflammatory agent, source of antibiotics and as a rich source of "high quality protein" have brought a revolution in the vermiculture studies.

About 4,400 different species of earthworms have been identified, and quite a few of them are versatile waste eaters and bio-degraders and several of them are bio-accumulators & bio-transformers of toxic chemicals.

THE BIOLOGY AND ECOLOGY OF EARTHWORMS

Earthworms are long, narrow, cylindrical, bilaterally symmetrical, segmented animals without bones. Usually the life span of an earthworm is about 3 to 7 years depending upon the type of species and the ecological situation. Earthworms harbor millions of "nitrogen-fixing" and "decomposer microbes" in their gut. They have "chemoreceptors" which aid in search of food. Their body contains 65% protein (70-80% high quality "lysine rich protein" on a dry weight basis), 14% fats, 14% carbohydrates and 3% ash [4-6].

Earthworms occur in diverse habitats specially those which are dark and moist. They can tolerate a temperature range between 5°C and 29°C. A temperature of 20°C to 25°C and moisture of 60-75% are optimum for good worm function. Earthworms multiply very rapidly. Studies indicate that they double their number at least every 60-70 days. Given the optimal conditions of moisture, temperature and feeding materials earthworms can multiply by 2^8 i.e. 256 worms every 6 months from a single individual. Each of the 256 worms multiplies in the same proportion to produce a huge biomass of worms in a short time. The total life-cycle of the worms is about 220 days. They produce 300-400 young ones within this life period [7]. Earthworms continue to grow throughout their life.

Earthworms can tolerate toxic chemicals in environment. After the Seveso chemical plant explosion in 1976 in Italy, when a vast area was contaminated with extremely toxic chemical like TCDD (2,3,7,8-tetrachlorodibenzo-p-dioxin) several fauna perished except for some species of the earthworms that survived. Earthworms which ingested TCDD contaminated soils were shown to bio-accumulate dioxin in their tissues and concentrate it on average 14.5 fold [8]. Earthworms are also tolerant to moderate salt salinity in soil, but some species like the tiger worms (Eisenia fetida) has been found to be highly salt tolerant. In a study made by [9] at the US Department of Energy it was found that replicates with a salt concentration of 15 g/kg of soil (nearly half as salty as seawater), survival ranged from 80%-100%.

TECHNOLOGIES FOR SUSTAINABLE DEVELOPMENT BY USE OF| EARTHWORMS

Following technologies for sustainable development with environmental protection can be envisaged by the use of useful earthworms species which promises to provide cheaper solutions to several social, economic, environmental & health problems plaguing the human society [3,10].

1) "THE VERMI-COMPOSTING TECHNOLOGY" for efficient management of municipal & industrial solid wastes (organics) by biodegradation & stabilization and converting them into useful resource (vermicompostnutritive biofertilizer);

2) "THE VERMI-FILTRATION TECHNOLOGY" for treatment of municipal and some industrial wastewater, their purification & disinfection for reuse;
3) "THE VERMI-REMEDIATION TECHNOLOGY" for cleaning up chemically contaminated sites (lands) while also improving their physical, chemical and biological properties for reuse;
4) "THE VERMI-AGRO-PRODUCTION TECHNOLOGY" for restoring & improving soil fertility to produce safe and chemical-free food for the society by the use of vermicompost & without recourse to the destructive agro-chemicals;
5) "THE VERMI INDUSTRIAL PRODUCTION TECHNOLOGY" for use of earthworms to produce some "bioactive compounds" for pharmaceutical industries and valuable raw materials to be used in rubber, lubricant, soaps, detergent & cosmetics industries and use of rich worm proteins as feed materials to promote fishery, dairy & poultry industries to produce more nutritive foods for the society.

Vermiculture technologies based on earthworms are self-promoted, self-regulated, self-improved & self-enhanced, low or no-energy requiring zero-waste technologies, easy to construct, operate and maintain. They excel all "bio-conversion", "bio-degradation" & "bio-production" technologies by the fact that they can utilize organics that otherwise cannot be utilized by others. They excel all "bio-treatment" technologies because they achieve greater utilization than the rate of destruction achieved by other technologies. They involve about 100-1000 times higher "value addition" than other biological technologies [11,12].

Technologies based on earthworms are also "environmentally & economically sustainable" as the worms are "highly renewable resources" regenerating at a rapid rate (by 2^8 i.e. 256 worms every 6 months from a single individual & each of the 256 worms multiplying in the same proportion) and the products are completely "biodegradable". The best part is that application of some vermiculture technologies are based on the by-products (worm biomass & vermicompost) generated in the operation of other technologies and therefore, more sustainable. It is like "killing several birds" in one shot.

THE VERMICOMPOSTING TECHNOLOGY FOR EFFICIENT MANAGEMENT OF SOLID WASTES: DIVERTING WASTE FROM LANDFILLS, AND RECOVERING RESOURCE FROM WASTE

Waste is a problem of the modern civilized society. We are facing the escalating socio-economic and environmental cost of dealing with current and future generation of mounting municipal solid wastes (MSW). A considerable portion of MSW consist of "Organic Wastes" that are "biodegradable" and can be vermicomposted into a highly "nutritive bio-fertilizer" 4-5 fold more powerful than conventional composts and even superior to chemical fertilizers for better crop growth and safe food production.

Another serious cause of concern today is the emission of greenhouse gases (GHG) methane (CH_4) & nitrous oxides (N_2O) resulting from the disposal of MSW either in the landfills or from their management by conventional composting systems. Molecule to molecule CH_4 is 21 times and N_2O is 310 times more powerful GHG than the CO_2. Millions of tons of MSW generated from the modern society are ending up in the landfills everyday, creating extraordinary economic and environmental problems for the local government to manage and monitor them (may be up to 30 years) for environmental safety (emission of GHG, toxic gases and leachate discharge into ground water). Construction of secured engineered landfills incurs 20-25 million U.S. dollars before the first load of waste is dumped in 2002-03. Over the past 5 years the cost of landfill disposal of waste has increased from $ 29 to $ 65 per ton of waste in Australia. During 2002-2003, waste management services within Australia cost $ 2458.2 million.

Waste degradation & composting by earthworms is proving to be economically & environmentally preferred technology over the conventional microbial degradation & composting technology as it is rapid and nearly odorless process, reducing composting time by more than half and the end product is both "disinfected" and "detoxified"

[13,14]. On an average, 2000 adult worms weigh 1 kg & one million worms approx. 1 ton. One million worms doubling every two months can become 64 million worms at the end of the year. Considering that each adult worm (particularly Eisinia fetida) consume waste organics equivalent to its own body weight every day, 64 million worms (weighing 64 tons) would consume 64 tons of waste every day and produce 30-32 tons of vermicompost per day at 40-50% conversion rate.

Community Wastes that can Be Salvaged by Earthworms

Waste eater earthworms can physically handle a wide variety of organic wastes from both municipal (domestic and commercial) and industrial (livestock, food processing and paper industries) streams [2,13,15].

Municipal Organic Wastes

The food waste from homes (both raw & cooked kitchen wastes—fruits and vegetables, grains & beans, coffee grounds, used tea leaves & bags, crushed egg shells) and restaurants & fried food wastes from fast-food outlets The garden wastes (leaves and grass clippings) also constitute an excellent feed stock for vermi-composting. Grass clippings (high carbon waste) require proper blending with nitrogenous wastes. The "sewage sludge" (biosolids) from the municipal wastewater also provide a good feedstock for the worms. The worms digest the sludge and convert a good part of it into vermi-compost. Paunch waste materials (gut contents of slaughtered ruminants) from abattoir also make good feedstock for earthworms [14,16].

Agriculture and Animal Husbandry Wastes

Farm wastes such as crop residues, dry leaves & grasses. Livestock rearing waste such as cattle dung, pig and chicken excreta makes excellent feedstock for earthworms.

Some Industrial Organic Wastes

Solid waste including the "wastewater sludge" from paper pulp and cardboard industry, food processing industries including brewery and distillery; vegetable oil factory, potato and corn chips manufacturing industry, sugarcane industry, aromatic oil extraction industry. Sericulture industry, logging and carpentry industry also offers excellent feed material for vermi-composting by earthworms [17,18].

Some Important Studies on Vermicomposting Technology

1) Reference [6] studied vermicomposting in great details and found that most earthworms consume, at the best, half their body weight of organics in the waste in a day. Eisenia fetida can consume organic matter at the rate equal to their body weight every day. Earthworm participation enhances natural biodegradation and decomposition of organic waste from 60 to 80% over the conventional aerobic & anaerobic composting. Given the optimum conditions of temperature (20-30°C) and moisture (60-70%), about 5 kg of worms (numbering approx. 10,000) can vermiprocess 1 ton of waste into vermicompost in just 30 days. Upon vermi-composting the volume of solid waste is significantly reduced from approximately 1 cum to 0.5 cum of vermi-compost indicating 50% conversion rate, the rest is converted into worm biomass.
2) References [18-22] studied the degradation and composting of "wastewater sludge" from paper pulp and cardboard industry, brewery and distillery, sericulture industry, vegetable oil factory, potato and corn chips manufacturing industry, sugarcane industry, aromatic oil extraction industry, logging and carpentry industry by earthworms. These organic wastes offer excellent feed materials for vermi-composting by earthworms. Reference [19] also studied the vermicomposting of waste from the mining industry which contains sulfur residues and creates disposal problems. They can also be fed to the worms mixed with organic matter. Optimum mixing ratio of the sulfur waste residues to the organic matter was 4%.

3) Reference [23] studied the vermicomposting of "fly-ash" from the coal power plants which is considered as a hazardous waste and poses serious disposal problem due to heavy metal contents. As it is also rich in nitrogen and microbial biomass it can be vermi-composted by earthworms. They found that 25% of fly-ash mixed with sisal green pulp, parthenium and green grass cuttings formed excellent feed for Eisenia fetida and the vermicompost was higher in NKP contents than other commercial manures. The earthworms ingest the heavy metals from the fly-ash while converting them into vermicompost.

4) Reference [24] successfully studied vermicomposting of "human excreta" (feces). It was completed in six months, with good physical texture, odourless and safe pathogen quality. Sawdust appeared to be the best covering material that can be used in vermicomposting toilets to produce compost with a good earthy smell, a crumbly texture and dark brown colour.

Our Studies on Vermicomposting Technology at Griffith University, Australia

1) Reference [25] studied the degradation and composting abilities of three species of earthworms on cattle dung, raw food wastes and garden wastes and found that the worm Eudrilus euginae was a better waste degrader followed by Eisenia fetida.

2) Reference [26] studied the vermicomposting of "sewage sludge" (biosolids). In 12 weeks study period there were no significant changes in the control. Most significant and rapid changes were observed in sludge which contained earthworms. Foul odor disappeared by week 2 and by week 12, the black and brittle sludge became a homogenous and porous mass of brown vermicast with light texture. Upon chemical analysis, the vermicomposted sludge was over 80% free of heavy metals cadmium (Cd) and lead (Pb) and almost completely free of any pathogens.

3) Reference [27] studied the efficiency of earthworms in degradation and vermicomposting of mixed food & garden wastes and compared with conventional aerobic composting without worms. Degradation of food wastes had started within hours (5%

after 24 hours) and were degraded 100% in just 15 days while garden wastes in 60 days. In wastes where the primary cellulosic materials were intact e.g. leaves and grasses, raw vegetables & fruits or where there are brittle calcium compounds e.g. egg shells, were degraded rather more slowly by the earthworms. In the conventional composting system without worms, maximum degradation of both food & garden wastes were only 35% even after 90 days.

Mechanism of Worm Action in Vermicomposting

Earthworms act as an aerator, grinder, crusher, chemical degrader and a biological stimulator and degrade waste by multiple actions.

Grinding Action

The waste feed materials ingested is finely ground (with the aid of stones in their muscular gizzard) into small particles to a size of 2-4 microns and passed on to the intestine for enzymatic actions. The gizzard and the intestine work as a "bioreactor";

Enzymatic Action

The gizzard and the intestine work as a "bioreactor". Worms secrete enzymes proteases, lipases, amylases, cellulases and chitinases in their gizzard and intestine which bring about rapid biochemical conversion of the cellulosic and the proteinaceous materials in the waste organics. They ingest the food materials, cull the harmful microorganisms, and deposit them mixed with minerals and beneficial microbes as "vermicasts" in the soil [28].

Worms Reinforce Decomposer Microbes and Act Synergistically

Worms promotes the growth of "beneficial decomposer microbes" (bacteria, actinomycetes & fungi) in waste biomass [28,29]. They hosts millions of decomposer microbes in their gut which is described as "little bacterial factory". They devour on microbes and excrete them out (many times more in number than they ingest) in soil along with nutrients nitrogen (N) and phosphorus (P) in their excreta. The nutrients N and P are further used by the microbes for multiplication and vigorous action [30]. Reference [31] showed that the number of bacteria and "actinomycetes" contained in the ingested material increased up to 1000 fold while passing through the gut. A population of worms numbering about 15,000 will in turn foster a microbial population of billions of millions. Earthworms and microbes act "symbiotically & synergistically" to accelerate and enhance the decomposition of the organic matter in the waste. It is the microorganisms that break down the cellulose in the food waste, grass clippings and the leaves from garden wastes [32].

Humification

The final process in vermi-processing and degradation of organic matter is the "humification" in which the large organic particles are converted into a complex amorphous colloid containing "phenolic" materials. About one-fourth of the organic matter is converted into humus. Humus is essential in soil for plant growth & survival. It takes several years in conventionally produced compost to generate "humus" through slow disintegration, whereas, earthworms excrete "humus" in its compost.

Advantages of Vermicomposting Technology

Salvage Human Wastes and Divert Huge MSW from Landfills

A big advantage of great economic & environmental significance is that production of vermicompost from "organic wastes" divert huge amount of MSW from ending up in the landfills thus also saving cost on waste disposal and reducing discharge of toxic leachate and gases and emission of greenhouse gases (GHG) which occurs from waste landfills.

Rapid and Efficient Process and Takes Half the Time than the Conventional System

Earthworms have the real potential to accelerate and enhance the natural biodegradation and decomposition of organic materials from 60 to 80% by promoting the growth of "beneficial decomposer aerobic bacteria" in the waste biomass. They hosts millions of decomposer microbes in their gut and also act as an aerator, grinder, crusher, chemical degrader and a biological stimulator [25]. As compared to the conventional composting systems, it takes nearly half the time to convert waste into vermicompost and the process becomes faster with time as the degrader worms and microbes multiply in number. Given the optimum conditions of temperature and moisture, about 1/2 kg of adult worms (approx. 1000) or 5 kg of worms (approx. 10,000) can vermicompost 10 kg of waste or 1 ton of waste respectively in just 30 days [6].

Earthworms Disinfect, Detoxify and Add Value to the End Products

Vermicomposting excels all conventional composting systems bythe fact that it can utilize organics that otherwise cannot be utilized by others and achieves greater "utilization" than the rate of "destruction" achieved by others and also involves about 100-1000 times higher "value addition" than other technologies [33]. Another matter of considerable significance is that the earthworms stabilize the organic residues in the waste removing any "harmful pathogens" and "toxic chemicals" from the compost. They partially "detoxify" and "disinfect" the end product which is nearly "odorless" [34]. Several studies have found that earthworms effectively bio-accumulate or biodegrade several organic and inorganic chemicals including "heavy metals", "organochlorine pesticide" and "polycyclic aromatic hydrocarbons" (PAHs) residues in the medium in which it inhabits [26]. The quality of vermicompost is significantly better, rich in key minerals & beneficial soil microbes. In fact in the conventional composting technology which is thermophilic (temperature rising up to 55°C) many beneficial microbes are killed and nutrient especially nitrogen is lost (due to gassing off of nitrogen). The end product is more homogenous, richer in "plant-available nutrients

& humus" and significantly low contaminants. They are "soft", "highly porous" with greater "water holding capacity" [1,11,35].

No or Low Energy Use in Vermi-composting Process

Normal microbial composting requires energy for aeration (constant turning of waste biomass and even for mechanical airflow) and sometimes for mechanical crushing of waste to achieve uniform particle size. Vermi-composting does not involve such use of energy. Earthworms aerate the system constantly by burrowing actions.

Generate No or Low Odor

Earthworms create aerobic conditions in the waste materials by their burrowing actions, inhibiting the action of anaerobic micro-organisms which release foulsmelling hydrogen sulfide and mercaptans.

Low Greenhouse Gas Emissions in MSW Disposal

High volumes of carbon dioxide (CO_2), methane (CH_4) and nitrous oxides (N_2O) is emitted from the conventional composting process especially in anaerobic conditions. Worms significantly increase the proportion of "aerobic to anaerobic decomposition" in the compost pile by "burrowing and aerating action" leaving very few anaerobic areas in the pile, and thus significantly reducing emission these gases and also volatile sulfur compounds. Analysis of vermicompost samples has shown generally higher levels of available nitrogen (N) as compared to the conventional compost samples made from similar feedstock. This implies that the vermicomposting process by worms is more efficient at retaining nitrogen (N) rather than releasing it as nitrous oxide (N_2O). Our study found that on average the anaerobic composting systems emitted the highest amount of CO_2 (2950 mg/m^2/hour) and CH_4 (9.54 mg/m^2/hour), while the aerobic systems (both with and without worms) emitted the least amount of CO_2 (880 mg/m^2/hour) and CH_4 (2.17 mg/m^2/hour). Vermicomposting systems had the "lowest emission" of N_2O which is most powerful GHG [2,36].

Global Movement for Vermicomposting MSW to Divert Waste from Landfills

Large scale vermicomposting of MSW including the "sewage sludge" on commercial scale is a movement going on to divert solid waste from ending up in the landfills [37]. Municipal councils and composting companies are also participating in vermicomposting business, composting all types of organic wastes on commercial scale and selling them to the farmers. This has dual benefits. Cutting cost on landfill disposal of waste while earning revenues from sale of worms & vermicompost [1,10,38]. First serious experiments for management of municipal/industrial organic wastes were established in Holland in 1970, and subsequently in England, and Canada. Later vermiculture were followed in USA, Italy, Philippines, Thailand, China, Korea, Japan, Brazil, France, Australia, Israel & Russia.

1) ***USA***: U.S. has some largest vermicomposting companies and plants in world and States are encouraging people for "backyard vermicomposting" to divert wastes from landfills [39]. The American Earthworm Company started a "vermi-composting farm" in 1978-79 with 500 t/month of vermicompost production [40]. A farm in LA rears 1,000,000 worms to treat 7.5 tons of garbage each month. Nearly 300 large-scale vermiculturist formed an "International Worms Growers Association" in 1997 and is having booming business. Vermicycle Organics produced 7.5 million pounds of vermicompost every year in high-tech greenhouses. Its sale of vermicompost grew by 500% in 2005. Vermitechnology Unlimited has doubled its business every year since 1991 [41,42].

US scientists are also searching for life-saving "vermimedicines" from the bioactive compounds in earthworms [43].

2) ***Canada***: Canada is also ahead in vermicomposting business on commercial scale for both "vermicompost" and "vermimeal" production. Large-scale vermicomposting plants have been installed at several places to vermicompost municipal and farm wastes and their use in agriculture [44]. An "Organic Agriculture Centre of Canada" has been established whose objective is to replace "Chemical Agriculture" by Vermiculture [38].

3) ***UK***: UK is also following US and Canada in promoting vermiculture mainly for waste management and to reduce the needs of "waste landfills". Large 1000 metric ton vermi-composting plants have been erected in Wales to compost diverse organic wastes [45].

4) ***France***: France is also promoting vermiculture on commercial scale to manage all its MSW and reduce the needs of landfills. About 20 tons of mixed household wastes are being vermi-composted everyday using 1000 to 2000 million red tiger worms (Eisenia andrei) [6].

5) ***New Zealand***: It is also a leading nation in vermiculture. The Envirofert Company of New Zealand is vermicomposting thousands of tons of green waste every year. They put the green waste first to a lengthy thermophilic cooking, and then to vermicomposting by worms after cooling. Cooking of green waste help destroy the weeds and pathogens which may come from the feces of pets in grasses. They claim that each worm eat the cooked green waste at least 8 times leaving an end product rich in key minerals, plant growth hormones, enzymes, and beneficial soil microbes. Envirofert is also planning to vermicompost approximately 40,000 tones of food wastes from homes, restaurants and food processing industries every year (www.envirofert.co.nz) [45,46].

6) ***Australia***: Vermicomposting is being done on large scale in Australia as a part of the "Urban Agriculture Development Program" utilizing the urban solid wastes [1]. The Sydney Waters in New South Wales have set up a vermiculture plant of 40 million worms to degrade up to 200 ton of urban wastes a week. The Gayndah Shire Council in Queensland, Australia, is vermi-composting over 600 tons of organic waste into valuable organic fertilizer (vermi-compost) and selling to the local farmers. Vermicomposting of sludge from the sewage and water treatment plants is being increasingly practiced in Australia and as a result it is saving over 13,000 cum of landfill space every year in Australia [47]. The Hobart City Council in Tasmania, vermicompost and stabilize about 66 cum of sewage sludge every week.

7) ***India***: India also launched vermicomposting program of MSW in the 1990s and and Bhawalkar Earthworms Research Institute (BERI) in Pune were among the pioneer institutions. Tata Energy

Research Institute (TERI) in Delhi is also doing commendable works. In recent years it is growing as a part of "sustainable non-chemical agriculture" program combined with "poverty eradication" program. Farmers are using vermicompost on large scale and a revolution is going on. Vermicomposting business has enhanced the lives of poor in India and generated self-employment opportunities for the unemployed. In several Indian villages NGO's are freely distributing cement tanks and 1000 worms and encouraging men and women to collect waste from villages and farmers, vermicompost them and sell both worms and vermicompost to the farmers. People are earning from Rupees 5 to 6 lakhs (Approx. AU $ 15-20 thousands) every year from sale of both worms and their vermicompost to the farmers. Mostly they use farm waste and also MSWs collected from streets and waste dumpsites [48,49].

Bihar, Karnataka, Tamil Nadu, Gujarat and Mahrashtra are leading states in vermiculture revolution. The Karnataka Compost Development Corporation established a first vermicomposting unit in the country to handle all municipal urban solid wastes and is producing 150 to 200 tons of vermicompost every day from city garbage [50]. She has listed several farmers whose life has been changed from a poor "farm labourer" to a "rich farmer" who embraced vermiculture.

8) ***Philippines***: Vermiculture and vermicomposting were introduced in the Philippines in the 1970s. Vermicompost is being used by farmers on large scale replacing the chemical fertilizers. Recently, commercial production of "vermimeal" from earthworms biomass has been started as a substitute to "fishmeal" for promoting fishery industries [51].

9) ***Argentina***: Vermiculture is an expanding business in Argentina especially for the development of rural and farming communities. "Worms Argentina" is a growing company which reports to be exporting "composting worms" on large scales to European, South American, Caribbean and Middle East nations. They are in high demands from Middle East countries for recycling of polluting dairy effluents [52].

10) ***China***: Vermiculture is a fast growing industry in China for the development of rural communities. It is in fact "revival" of the traditional culture practiced by ancient medicinemen who used

earthworms for treatment of several diseases. Earthworms are now being used for vermicomposting of "waste", promoting "organic farming" and for the development of "vermi-medicines" and "nutritive vermimeals". A dietary supplement in the name of PLASMIN is being marketed in China [42,53, 54].

11) ***Russia***: Vermiculture is being promoted on large scale in Russia for management of MSW and development of life-saving "vermi-medicines" from the worms for treatment of human diseases for which conventional medicine do not have an answer. Scientists have developed a special breed of the versatile species Eisenia fetida which can tolerate and survive in cold climates [55].

12) ***Japan***: Japan is also promoting vermiculture since 1970s mainly for management of MSW and production of worm biomass to isolate "bioactive compounds" for production of "vermi-medicines" [12,56].

THE VERMIFILTRATION TECHNOLOGY FOR WASTEWATER TREATMENT, DISINFECTION & DETOXIFICATION FOR REUSE

We are also facing the escalating socio-economic and environmental cost of dealing with current and future generation of mounting municipal and industrial wastewater. Over 80% of the potable water used by society return as wastewater. Conventional treatment results into formation of "sludge" which requires safe disposal in "secured landfills" at additional cost.

Vermifiltration of wastewater using waste eater earthworms is a newly conceived novel technology with several advantages over the conventional systems. Earthworms body work as a "biofilter" and they have been found to remove the 5 days biological oxygen demand (BOD_5) by over 90%, chemical oxygen demand (COD) by 80-90%, total dissolved solids (TDS) by 90-92% and the total suspended solids (TSS) by 90-95% from wastewater by the general mechanism of "ingestion" and biodegradation of organic wastes and also by their "absorption" through body walls. Suspended solids are trapped on top of the vermifilter and processed by earthworms and fed to the soil microbes immobilized in the vermifilter. Worms also remove chemicals including heavy metals and pathogens from treated wastewater [57] and the treated water becomes fit for "reuse" in non-potable purposes.

Some Important Studies on Vermifiltration Technology

1) Reference [58] studied the use of earthworm for the management of effluents from intensively housed livestock which contain very heavy loads of BOD, TDSS and nutrients nitrogen (N) and phosphorus (P). The worms produced clean effluents and also nutrient rich vermicompost.
2) Reference [59] studied the treatment of domestic wastewater using vermifilter beds and concluded that worms can reduce BOD and COD loads as well as the TDSS (total dissolved and suspended solids) significantly by more than 70-80%.
3) Reference [60] studied the vermifiltration of municipal wastewater in a pilot plant for treating wastewater of 1000 inhabitants and found that the BOD load was removed by 99%, TSS by 95%, VSS (volatile suspended solids) by 96%, nitrogen (N) by 89% and phosphorus (P) by 70%. The vermifilter bed was prepared of stones at the bottom and sawdust above with 20-30 cm humus at the top in which 5000-10,000 earthworms (Eisenia andrea) per square meter was released. E. coli (M.P.N.) was removed by 1000 fold. Such systems allowed to treat 1000 L/m^2 of wastewater per day. They have commercialized and patented the technology in Chile.
4) A pilot study on vermifiltration of sewage was made by [61] at Shanghai Quyang Wastewater Treatment Facility in China. The earthworm bed which was 1 m (long) × 1 m (wide) × 1.6 m (high), was composed of granular materials and earthworms. The worm's number was kept at about 8000 worms/sqm. The average chemical oxygen demand (COD) value of raw sewage used was 408.8 mg/L that of 5 days biological oxygen demand (BOD_5) was 297 mg/L that of suspended solids (SS) was 186.5 mg/L. The hydraulic retention time varied from 6 to 9 hours and the hydraulic loading from 2.0 to 3.0 m^3/ (m^2.d) of sewage. The

removal efficiency of COD ranged between 81-86%, the BOD_5 between 91-98%, and the SS between 97-98%.

Our Studies on Vermifiltration Technology at Griffith University, Australia

1) Reference [62] studied the vermifiltration of sewage obtained from the Oxley Wastewater Treatment Plant in Brisbane, Australia. Results showed that the earthworms removed BOD (BOD_5) loads of sewage by over 99% at hydraulic retention time (HRT) of 1-2 hours. Average COD removed from the sewage was over 50%. Although the COD removal by vermifiltration system was not significant like BOD, it was still higher than the value of COD removed by the control system without worms. This clearly signifies the role of earthworms in the degradation of "complex chemicals" in wastewater to reduce the Chemical Oxygen Demand. Earthworms also removed the total suspended solids (TSS) from the sewage by over 90%. More significant was that there was "no formation of sewage sludge".

2) Reference [63] also studied the vermifiltration of brewery and milk dairy wastewaters in Brisbane which have very high BOD_5 and TSS loadings e.g. 6780 mg/L & 682 mg/L respectively from brewery and 1,39,200 mg/L & 3,60,00 mg/L respectively from the dairy industry. Earthworms removed the high BOD_5 loads by 99% in both cases and TSS by over 98%. But the hydraulic retention times (HRTs) in case of brewery wastewater was 3-4 hours and 6-10 hours for the dairy wastewater.

   An important observation was that the control system (devoid of earthworms) frequently got choked. The organic solids in the wastewater accumulated as peat in the soil layer and also attracted heavy "fungal infection". It became un-operational after sometimes. In the vermifiltration system the earthworms constantly fed upon the solids and the fungus and never allowed the system to be choked and become un-operational.

3) Currently Reference [64] are working on vermifiltration of wastewater from "fruit juice processing industry" in Brisbane. The fruit juice wastewater contain very high BOD, COD, TSS and TDS loads. The initial 5 days biological oxygen demand (BOD_5)

was reduced from 1340 ppm to 3 ppm (99.77%), chemical oxygen demand (COD) from 2730 ppm to 112 ppm (95.89%), total suspended solids (TSS) from 190 ppm to 16 ppm (91.57%), total dissolved solids (TDS) from 440 ppm to 12 ppm (97.27%), and the turbidity from 130 NTU (naphelometric units) to 6 NTU (95.38%). The vermifiltration system (vermifilter bed) contained a 4 inches layer of vermicompost (humus) and on the top of it a 6 inches layer of soil containing 450-500 "adult healthy earthworms" (Eisenia fetida) @ 5000 worms per cubic meter of soil.

The Mechanism of Worm Action in Vermifiltration

The twin processes of microbial stimulation & biodegradation, and the enzymatic degradation of waste solids by worms simultaneously work in the vermifiltration system. Vermifilters provide a high specific area—up to 800 sq m/g and voidage up to 60%. Suspended solids are trapped on top of the vermifilter and processed by earthworms and fed to the soil microbes immobilized in the vermifilter.

Intensification of soil processes and aeration by the earthworms enable the soil stabilization and filtration system to become effective and smaller in size. Earthworms intensify the organic loadings of wastewater in the vermifilter soil bed by the fact that it granulates the clay particles thus increasing the "hydraulic conductivity" of the system. They also grind the silt and sand particles, thus giving high total specific surface area, which enhances the ability to "adsorb" the organic and inorganic from the wastewater passing through it.

Earthworms vermicompost in the vermifilter beds provides wonderful sites for "adsorption" of heavy metals and pollutants in wastewater due to presence of "lignin" contents. The vermicast in the soil bed also offers excellent hydraulic conductivity of sand (being porous like sand) and also high adsorption power of clay [48].

Advantages of Vermifiltration Technology over the Conventional Wastewater Treatment Technologies

Low Energy Requirement

Vermi-filtration of wastewater is low energy & efficient system and has distinct advantage over all the conventional wastewater treatment systems—the "Activated Sludge Process", "Trickling Filters" and "Rotating Biological Contactors" which are highly energy intensive, costly to install and operate and do not generate any income. In the vermifilter process there is 100% capture of organic materials, the capital and operating costs are less, and there is high value added end products e.g. vermifiltered "nutrient rich" water with potential for farm irrigation and vermicompost retrieved from the vermifiltered beds.

No Formation of Sewage Sludge and No Foul Odor

Since the conventional technologies are mostly the flow-processes and have finite hydraulic retention time (HRT) it always results into a "residual stream" of complex organics and heavy metals (while only the simple organics are consumed by decomposer microbes) in the form of "sludge". This plagues most municipal councils in world as the sludge is a "biohazard" and requires safe landfill disposal at high cost. The greatest advantage of vermifiltration system is that there is no formation of "sewage sludge" [65]. The worms decompose the organics in the wastewater and also devour the solids (which forms the sludge) synchronously. There is no foul odor as the earthworms arrests rotting and decay of all putrescible matters in the wastewater and the sludge. In all developed nations a "worm farm" has become a necessity in all wastewater & water treatment plants to resolve the sludge problems. Earthworms feed readily upon the sludge components, rapidly convert them into vermicompost, reduce the pathogens to safe levels and ingest the heavy metals [26].

Detoxified and Disinfected Treated Water for Reuse

Vermifiltered wastewater is free of pathogens and toxic chemicals (heavy metals & endocrine disrupting chemicals) and suitable for "reuse" as water for farm irrigation. & other non-potable uses. The worms devour on all the pathogens (bacteria, fungus, protozoa & nematodes) in the medium in which they inhabit. They have the capacity to bio-accumulate high concentrations of toxic chemicals in their tissues and the resulting wastewater becomes almost chemical-free. Earthworms have also been reported to bio-accumulate "endocrine disrupting chemicals" (EDCs) from sewage which otherwise is not removed by our conventional sewage treatment plants (STPs). Reference [66] have reported significantly high concentrations of EDCs (dibutylphthalate, dioctylphthalate, bisphenol-A and 17 b-estrdiol) in tissues of earthworms (E. fetida) living in sewage percolating filter beds and also in garden soil.

Vermifiltration Technology: Destined to Become a Global Movement

Due to its simplicity and cost-effectiveness vermifiltration of both municipal and industrial wastewater is destined to become a global movement. In Chile, over 100 sewage treatment plants of different sizes, going from individual houses to plants for 12,000 persons and bigger plants for industries are already working. It has been introduced on commercial scale in Mexico and Venezuela [60]. India and Brazil is also introducing the technology on commercial scale. Some companies in Pune (India) have already started pilot plants.

If a vermifilter bed of 0.3 cum soil is prepared with approximately 5000 worms (over 2.5 kg) to start with, it can easily treat 950-1000 L of domestic wastewater/ sewage generated by (on an average) a family of 4 people with average BOD value ranging between 300-400 mg/L, COD 100-300 mg/L, TSS, 300-350 mg/L everyday with hydraulic retention time (HRT) of the wastewater in the vermifilter bed being approximately 1-2 hours. Given that the worms multiply and double its number in at least every 60 days under ideal conditions of temperature and moisture, even starting with this number of earthworms a huge population (biomass) of worms with robust vermi-filtration system can be established quickly within few months which will be able to treat greater amount of wastewater generated in the family. An important consideration is the peak hour wastewater generation which is usually very high and may not comply with the required HRT (1-2 hrs) which is very critical for sewage treatment by vermi-filtration system. To allow 1-2 hrs HRT in the vermifilter bed an onsite domestic wastewater storage facility will be required from where the discharge of wastewater to the vermifilter tank can be slowly regulated through flow control.

VERMIREMEDIATION TECHNOLOGY FOR CLEANUP OF CHEMICALLY CONTAMINATED LANDS AND SOIL FOR RE-USE & RE-DEVELOPMENT

Large tract of arable land is being chemically contaminated due to mining activities, heavy use of agrochemicals in farmlands, landfill disposal of toxic wastes and other developmental activities like oil and gas drilling. Traditionally, remediation of chemically contaminated soils involves "off-site" management by excavating and subsequent disposal by burial in secured landfills. This method of remediation is very costly affair and merely shifts the contamination problem elsewhere. Additionally, this involves great risk of environmental hazard while the contaminated soils are being transported and "migration of contaminants" from landfills into adjacent lands and water bodies by leaching. Soil washing for removing inorganic contaminants from soil is another alternative to landfill burial, but this technique produce a "residue" with very high metal contents which requires further treatment or burial.

Since the late 1980s, after the chemical and mechanical treatments of lands and water bodies and thermal treatment (incineration) of hazardous wastes proved economically and environmentally unsustainable, focus shifted towards the biological methods which are cost-effective as well as environmentally sustainable and also socially acceptable.

Vermiremediation (using chemical tolerant earthworm species) is emerging as a low-cost and convenient technology for cleaning up the chemically polluted/contaminated sites/lands in world. Earthworms have been used for land recovery, reclamation and rehabilitation of suboptimal soils such as poor mineral soils, polder soils, open cast mining sites, closed landfill sites and cutover peat [67,68].

Some Important Studies on Vermiremediation Technology

Studies on Removal of Heavy Metals

Reference [69] studied that earthworms can bio-accumulate high concentrations of heavy metals like cadmium (Cd), mercury (Hg), lead (Pb) copper (Cu), manganese (Mn), calcium (Ca), iron (Fe) and zinc (Zn) in their tissues without affecting their physiology and this particularly when the metals are mostly non-bioavailable. They can particularly ingest and accumulate extremely high amounts of zinc (Zn), lead (Pb) and cadmium (Cd). Cadmium levels up to 100 mg per kg dry weight have been found in tissues. Ireland (1983) reported that the earthworms species Lumbricus terrestris can bio-accumulate in their tissues 90-180 mg lead (Pb)/gm of dry weight, while L. rubellus and D. rubida it was 2600 mg /gm and 7600 mg/gm of dry weight respectively. Zinc (Zn), manganese (Mn), and iron (Fe) were shown to be excreted through the calciferous glands of earthworms. Contreras-Ramos et al. (2005) also confirmed that the earthworms reduced the concentrations of chromium (Cr), copper (Cu), zinc (Zn) and lead (Pb) in the vermicomposted sludge (biosolids) below the limits set by the USEPA in 60 days.

Studies on Removal of Polycyclic Aromatic Hydrocarbons (PAH's)

PAHs are priority pollutants and cause great concern with respect to human health and environment. They are inherently "recalcitrant hydrocarbons", and the higher molecular weight PAHs are very difficult to remediate. Reference [70] studied the influence of earthworms species L. rubellus on the disappearance of spiked PAHs phananthrene & fluoranthene (100 μg/kg of soil) and found that the losses of both PAHs occurred at a faster rate in soils with earthworms, than the soil without worms. After 56 days (8 weeks), 86% of the phenanthrene was removed. Reference [71] studied the uptake of three PAHs viz. phenanthrene, anthracene and benzo(a)pyrene at different concentrations by E. fetida and found that the concentration of anthracene decreased by 2-fold after addition of earthworms, benzo(a)pyrene decreased by 1.4-fold and phenanthrene was completely removed (100%) by earthworms

Studies on Removal of Petroleum and Crude Oil Hydrocarbons

References [72] & [73] studied earthworm species E. fetida with varying organic wastes to an oil contaminated soil and found that worms significantly degraded and decreased oil contents in comparison to the control. Reference [74] also studied the use of earthworms E. fetida and vermicomposting in the treatment of high molecular weight hydrocarbons "asphaltens" from the Prestige Oil Spill. Earthworms mineralized the asphaltens thus eliminating it from the system.

Studies on Removal of Agrochemicals

Studies indicate that the earthworms bio-accumulate or biodegrade "organochlorine pesticide" and "polycyclic aromatic hydrocarbons" (PAHs) residues in the medium in which it lives. References [75-78] found that the worm vermicasts sorbed higher amount of herbicides from the contaminated soil than the control soil due to the higher levels of organic carbon & more finer size of fractions in worm worked contaminated soils. Reference [79] found that due to earthworm burrowing actions, a greater degree of bound pesticides residues in soil was released as compared to those without worms. Earthworms restricted the formation of bound fraction of pesticides and also enhanced the release and mineralization of bound pesticides residues.

Studies on Removal of Polychlorinated Biphenyls (PCBs)

PCBs are a group of oily, colorless, organic fluids belonging to the same chemical family as the pesticide DDT. PCBs are categorized as unusually toxic and "persistent organic pollutant" (POPs). Reference [80] found that PCB contaminated soil treated with earthworms resulted in significantly greater PCB losses (average 52%) when compared to the soil without earthworm treatment which was 41%.

Our Studies on Vermiremediation Technology at Griffith University, Australia

Reference [81] studied the remedial action of earthworms on PAHs contaminated soils obtained from a former gas works site in Brisbane where gas was being produced from coal. The initial concentration of total PAHs compounds in the soil at site was greater than 11, 820 mg/kg of soil. The legislative requirements for PAHs concentration in soil in Australia is only 100 mg/kg for industrial sites and 20 mg/kg for residential sites. Results showed that the earthworms could remove nearly 80% of the PAHs as compared to just 47% & 21% where it was not used and only microbial degradation occurred. This was just in 12 weeks study period. It could have removed by 100% in another few weeks. More significant was that the worm added soil became odor-free of chemicals in few days and were more soft and porous in texture.

Mechanism of Worm Action in Vermiremediation

Earthworms uptake chemicals from the soil through passive "absorption" of the dissolved fraction through the moist "body wall' in the interstitial water and also by mouth and "intestinal uptake" while the soil passes through the gut. Earthworms apparently possess a number of mechanisms for uptake, immobilization and excretion of heavy metals and other chemicals. They either "bio-transform" or "biodegrade" the chemical contaminants rendering them harmless in their bodies. Some metals are bound by a protein called "metallothioneins" found in earthworms which has very high capacity to bind metals. The chloragogen cells in earthworms appears to mainly accumulate heavy metals absorbed by the gut and their immobilization in the small spheroidal chloragosomes and debris vesicles that the cells contain [70,76] found that earthworms biodegrade organic contaminants like phthalate, phenanthrene and fluoranthene.

Advantages of Vermiremediation Technology over the Mechanical and Chemical Treatment of Contaminated Sites

The greatest advantage of the vermiremediation technology is that it is "on-site" treatment and there are no additional problems of "earth-cutting", "excavation" and "transportation" of contaminated soils to the landfills or to the treatment sites incurring additional economic and environmental cost. Vermiremediation would cost about $ 500-1000 per hectare of land as compared to $ 10,000- 15,000 per hectare by mechanical excavation of contaminated soil & its landfill disposal.

Significantly, vermiremediation leads to total improvement in the quality of soil and land where the worms inhabit. Earthworms significantly contribute as soil conditioner to improve the physical, chemical as well as the biological properties of the soil and its nutritive value. They swallow large amount of soil every day, grind them in their gizzard and digest them in their intestine with aid of enzymes. Only 5-10 percent of the digested and ingested material is absorbed into the body and the rest is excreted out in soil in the form of fine mucus coated granular aggregates called "vermicastings" which are rich in NKP (nitrates, phosphates and potash), micronutrients and beneficial soil microbes including the "nitrogen fixers" and "mycorrhizal fungus".

Of considerable economic and environmental significance is that the worm feed used in vermiremediation process is necessarily an "organic waste" product. This means that it would also lead to reuse and recycling of vast amount of organic wastes which otherwise end up in landfills for disposal at high cost. And what is of still greater economic and environmental significance is that the polluted land is not only "cleaned-up" but also "improved in quality". The soil becomes lighter and porous rich in biological activities and the productivity is increased to several times. During the vermi-remediation process of soil, the population of earthworms increases significantly benefiting the soil in several ways. A "wasteland" is transformed into "wonderland". Earthworms are in fact regarded as "biological indicator" of good fertile soil and land.

Vermiremediation Technology Destined to Become a Global Movement

Vermiremediation by commercial vermiculture in U.K. "Land Reclamation and Improvements Programs" has become an established technology for long-term soil decontamination, improvement & maintenance, without earth-cutting, soil excavation and use of chemicals". U.S., Australia and other developed nations are also following [82,83].

VERMI-AGROPRODUCTION TECHNOLOGY FOR SUSTAINABLE AGRICULTURE AND PRODUCTION OF SAFE ORGANIC FOOD FOR SOCIETY

Vermi-agroproduction technology promises to usher in the "Second Green Revolution" by completely replacing the destructive agro-chemicals which did more harm than good to both the farmers and their farmland during the "First Green Revolution" of the 1950-60's. Studies indicate that vermicompost is a wonderful growth promoter and at least 4 times more nutritive than the conventional composts and gives 30-40% higher yield of crops over chemical fertilizers [84,85]. In Argentina, farmers consider it to be seven (7) times richer than conventional composts in nutrients and growth promoting values [52]. The "humic acid" in vermicompost (excreted by worms) stimulate plant growth even in small amount [86].

Earthworms restore & improve soil fertility and boost crop productivity by the use of their excreta (vermicast). They excrete beneficial soil microbes, and secrete polysaccharides, proteins and other nitrogenous compounds into the soil [87]. They promote soil fragmentation and aeration, and bring about "soil turning" and dispersion in farmlands. Worm activity can increase air-soil volume from 8-30%. One acre of land can contain up to 3 million earthworms the activities of which can bring up to 8-10 tons of "top soil" to the surface (in the form of vermicast) every year. Presence of worms improves water penetration in compacted soils by 50%. U.S. study indicates that 10,000 worms in a farm plot provides the same benefit as three farmers working 8 hours in shift all year round with 10 tons of manure applied in the plot [42]. Indian study showed that an earthworm population of 0.2-1.0 million per hectare of farmlands can be established within a short period of three months. On an average 12 tons/hectare/ year of soil or organic matter is ingested by earthworms, leading to upturning of 18 tons of soil/year, and the world over at this rate it may mean a 2 inches of fertile humus layer over the globe [48].

Some Important Studies on Vermi-Agroproduction Technology

1) Reference [85] studied the agronomic impacts of vermicompost and found that it consistently improved seed germination, enhanced seedling growth and development, and increased plant productivity much more than would be possible from the mere conversion of mineral nutrients into plant-available forms. The growth responses of plants from vermicompost appears more like "hormone-induced activity" associated with the high levels of nutrients, humic acids and humates in vermicompost rather than boosted by high levels of plant-available nutrients.
2) Studies made by [88] at CSIRO Australia found that the earthworms can increase growth of wheat crops by 39%, grain yield by 35%, lift protein value of the grain by 12% & fight crop diseases. Reference [89] also studied that earthworms & its vermicast improve the growth and yield of wheat by more than 40%.
3) Reference [90] studied the agronomic impacts of vermicompost and inorganic (chemical) fertilizers on strawberries when applied separately and also in combination. The "yield" of marketable strawberries and the "weight" of the "largest fruit" was greater on

plants in plots grown on vermicompost as compared to inorganic fertilizers. Also, farm soils applied with vermicompost had significantly greater "microbial biomass" than the one applied with inorganic fertilizers.

4) Reference [91] studied the agronomic impact of vermicompost on cherries and found that it increased yield of "cherries" for three (3) years after "single application" inferring that use of vermicompost in soil builds up fertility and restore its vitality for long time and its further use can be reduced to a minimum after some years of application in farms.

5) Reference [92] reported two-fold yield of grapes by vermicompost as compared to chemical fertilizers.

Our Studies on Vermi-Agroproduction Technology at Griffith University, Australia

1) Reference [93] studied the growth impacts of earthworms and their vermicompost on potted corn crops and compared with chemical fertilizers. Vermicompost with earthworms in soil achieved excellent growth over chemical fertilizers. While the plants on chemicals grew only 5 cm in 7 weeks, those on vermicompost with worms grew 15 cm within the same period. Corn plants with worms & vermicompost also attained maturity (appearance of male & female reproductive organs) very fast. Another significant finding was that plants on vermicompost demanded "less water" for irrigation.

2) Reference [93] also studied the growth impacts of earthworms with vermicompost on potted wheat plants and compared with chemical fertilizers & conventional compost (cow manure). Wheat crops on vermicompost with worms maintained very good growth from the very beginning & achieved maturity very fast. The striking rates of seed germination were very high, nearly 48 hours (2 days) ahead of others and the numbers of seed germinated were also high by nearly 20%. Plants were greener and healthier over others, with large numbers of tillers & long seed ears at maturity. Seeds were healthy and nearly 35-40% more as compared to plants on chemical fertilizers. What they achieved in just 5 weeks was achieved by others in 10 weeks.

More significant was that the pot soil with vermicompost was very soft & porous and retained more moisture. Pot soil with chemical fertilizers was hard and demanded more water frequently.

3) Reference [93] also studied the growth impacts of vermicompost on farmed wheat crops and compared it with conventional cattle dung compost and chemical fertilizers. Exclusive application of vermicompost@25 quintal/ha boosted yield 18% higher over the chemical fertilizers (NPK: 120:40:60). On conventional compost applied @ 100 Q/ha (4 times more than vermicompost) the yield was 17% less than that on vermicompost. The requirement of irrigation was also reduced in vermicompost applied farm plots by 30-40%. Test results indicated better availability of essential micronutrients and useful microbes in vermicompost applied soils. Most remarkable was the significantly reduced (nearly 75%) incidences of "pest & disease attack" on vermicompost grown crops.

Vermiculture Technology can Reduce Soil Salinity & Improve Fertility of Sodic Soils

Studies indicate that Esinea fetida can tolerate soils nearly half as salty as seawater i.e. 15 gm/kg of soil and also improve its biology and chemistry. (Average seawater salinity is around 35 g/L). Farmers at Phaltan in Satara district of Maharashtra, India, applied live earthworms to their sugarcane crop grown on saline soils irrigated by saline ground water. The yield was 125 tones/hectare of sugarcane and there was marked improvement in soil chemistry. Within a year there was 37% more nitrogen, 66% more phosphates and 10% more potash. The chloride content was less by 46% [93].

Reference [94] studied the production of potato (Solanum tuberosum) by application of vermicompost in a reclaimed sodic soil in India. With good potato growth the sodicity (ESP) of the soil was also reduced from initial 96.74 to 73.68 in just about 12 weeks. The average available nitrogen (N) content of the soil increased from initial 336.00 kg/ha to 829.33 kg/ha.

Advantages of Vermi-Agroproduction Technology

Can Replace Destructive & Costly Chemical Fertilizers from Farm Production

Vermicompost has potential to replace the destructive chemical fertilizers from farm production. It can alone produce food over 30-40% higher than those produced by chemical fertilizers. It is at least 75% cheaper than the chemical fertilizers which are produced in factories from vanishing petroleum products generating huge waste & pollution.

Produce Nutritive, Chemical-free Farm Products with Greater Storage Value

The biggest advantage of great social significance is that the food produced is completely organic "safe & chemical-free". Use of vermicompost enhances size, color, smell, taste, flavour and keeping quality (storage value) of flowers, fruits, vegetables and food grains.

Restore Natural Fertility of Farmland Soil

Upon successive years of application, vermicompost build-up the soils "natural fertility" improving its total physical (porous), chemical (rich in nutrients) and biological (beneficial soil microbes) properties. It also regenerates a rich population of worms in the farm soil from the cocoons which further help improve soil fertility and subsequently lesser amount of vermicompost is required to maintain a good yield and productivity. On the contrary, with the continued application of chemical fertilizers over the years the "natural fertility of soil is destroyed" and it becomes "addict". Subsequently greater amount of chemicals are required to maintain the same yield & productivity of previous years.

Reduces Water for Farm Irrigation

Vermicompost has very "high porosity", "aeration", "drainage" and "water holding capacity" and thus its application in soil reduces the requirement of water for irrigation by 30-40%.

Kills Pests without Pesticides

Another big advantage of great social & environmental significance is that vermicompost "suppress plant disease" in crops and inhibit the soil-born fungal diseases. In field trials with pepper, tomatoes, strawberries and grapes significant suppression of plant-parasitic nematodes has been found. There is also significant decrease in arthropods (aphids, buds, mealy bug, spider mite) populations with 20% and 40% vermicompost additions [95]. Humus in vermicast extracts "toxins", "harmful fungi & bacteria" from soil & protects plants. Actinomycetes in vermicast induces "biological resistance" in plants against pests & diseases. As such use of vermicompost significantly reduces the need for "chemical pesticides". Our studies indicated over 75%.

The Global Movement for VAPT to Replace the Destructive Chemical Fertilizer

Worldwide farmers are desperate to get rid of the vicious circle of the use of chemical fertilizers as their cost have been constantly rising and also the amount of chemicals used per hectare has been steadily increasing over the years to maintain the yield & productivity of previous years. Nearly 3-4 times of agro-chemicals are now being used per hectare what was used in the 1960s. In Australia, the cost of MAP fertilizer has risen from AU $ 530.00 to AU $ 1500.00 per ton since 2006. So is the story everywhere in world because the chemical fertilizers are produced from "vanishing resources" of earth. Farmers urgently need a sustainable alternative which is both economical and also productive while also maintaining soil health & fertility. The new concept is "Ecological Agriculture" which is by definition different from "Organic Farming" that was focused mainly on production of chemical-free foods. Ecological agriculture emphasize on total protection of food, farm & human ecosystems while improving soil fertility & development of secondary source of income for the farmers. UN has also endorsed it. Vermiculture technology provides the best answer for ecological agriculture which is synonymous with "sustainable agriculture".

THE VERMI-INDUSTRIAL PRODUCTION TECHNOLOGY FOR PRODUCING RAW MATERIALS FOR PHARMACEUTICAL & CONSUMER INDUSTRIES

Raw Materials for Rubber, Lubricant, Detergent, Soaps and Cosmetic Industries

Some biological compounds from earthworms are also finding industrial applications. Being "biodegradable" they are environmentally friendly and sustainable. Stearic acid found in earthworms is a long chain saturated fatty acid and is widely used as "lubricant" and as an "additive" in industrial preparations. It is used in the manufacture of metallic stearates, pharmaceuticals soaps, cosmetics and food packaging. It is also used as a "softner", "accelerator activator" and "dispersing agents" in rubbers. Industrial applications of lauric acid and its derivatives are as "alkyd resins", "wetting agents", a "rubber accelerator" and "softner" and in the manufacture of "detergents" and "insecticides" [54,96]. Worms are also finding new uses as a source of "collagen" for pharmaceutical industries.

Nutritive Feed Materials for Poultry, Dairy and Fishery Industries

Earthworms are rich in high quality protein (65%) and are "complete protein" with all essential amino acids. There is 70-80% high quality "lysine" and "methionine". Glumatic acid, leucine, lysine & arginine are higher than in fish meals. Tryptophan is 4 times higher than in blood powder and 7 times higher than in cow liver. Worms are also rich in Vitamins A & B. There is 0.25 mg of Vitamin B_1 and 2.3 mg of Vitamin B_2 in each 100 gm of earthworms. Vitamin D accounts for 0.04-0.073% of earthworms wet weight. Thus worms are wonderful probiotic feed for fish, cattle and poultry industry. They are being used as "additives" to produce "pellet feeds" in the USA, Canada and Japan [42].

As earthworm protein is complete with 8-9 essential amino acids especially with the tasty "glutamic acid" it can be used for human beings as well. Worm protein is higher than in any meat products with about 2% lower fats than in meats and ideal for human consumption [51].

Bioactive Compounds for Pharmaceutical Industries to Produce Life Saving Medicines

A great news for the world vermitech scientists appeared in Philippines "News Today" on November 25, 2005 telling "Earthworms can help dissolve blood clots for stroke patients" [97]. In the last 10 years, a number of earthworm's "clot-dissolving", "lytic" and "immune boosting" compounds have been isolated and tested clinically [98,99]. Current researches made in Canada, China, Japan and other countries on the identification, isolation and synthesis of some "bioactive compounds" from earthworms (Lumbricus rubellus & Eisenia fetida) with potential medicinal values for treatment of heart diseases have brought revolution in the vermiculture studies [12,100,101]. Some of these compounds have been found to be enzymes exhibiting "anti-blood clotting" effects. Oral administration of earthworms powder & enzymes were found to be effective in treating "thrombotic diseases", "arthritis", "diabetes mellitus", "pulmonary heart disease", "lowering blood pressure", "epilepsy", "schizophrenia", "mumps", "exzema", "chronic lumbago", "anemia", "vertigo" and "digestive ulcer" [44,102,103]. Scientist have also isolated "bronchial dilating" substance from earthworms.

Researchers at Quinghua University, China has extracted 4 valuable medicinal compounds from earthworms —a large molecular compound which has "anti-carcinogenic" effects [104]; medium molecular compound which has "anti-thrombosis" & "thrombus dissolution" effects [105]; a small molecular compounds which contain 17 kinds of amino acids, polymers, trace elements and vitamins; and a 4th product which can cure burns and scalds [42]. Reference [102] also extracted enzymes lumbritin, lumbrofebrin, terrestrolumbrolysin and "lumbrokinase" enzymes from Lumbricus rubellus useful in thrombolytic therapy.

The coelomic fluid of earthworms have been reported to have anti-pathogenic activities and are good biological compound for the production of "antibiotics" [34]. Several fatty acids have been isolated from earthworms. Important among them are "lauric acid" which are known for its "anti-microbial" properties. It is a precursor to "monolaurin" which is a more powerful "anti-microbial" agent that has potential to fight lipid-coated RNA and DNA viruses, several pathogenic Gram-positive bacteria, yeasts and various pathogenic protozoa [54]. Peptide "lumbricin I" isolated from L. lumbricus also exhibits antimicrobial activity against both Gram positive and Gram negative bacteria as well as fungi.

CONCLUSIONS & REMARKS

Vermiculture technologies for waste and land management and for improving soil fertility to promote crop productivity and production of valuable bioactive compounds of great medicinal values has grown considerably in recent years all over the world and has been scientifically improved [106]. It is like getting "gold from garbage" (highly nutritive biofertilizer) by vermi-composting technology; "silver from sewage" (disinfected & detoxified water for reuse in agriculture & industries) by vermi-filtration technology; "converting a wasteland (chemically contaminated lands) into wonderland" (fertile land) by vermi-remediation technology; harvesting "green gold" (food crops) by using "black gold" (vermicompost) by agro-production technology; creating a "worm factory" to produce medicines & materials for societal use. The three versatile species E. fetida, E. euginae and P. excavatus performing wide social, economic & environmental functions occur almost everywhere.

The vermi-composting & vermi-agro-production technologies can together maintain the "global human sustainability cycle" & "circular economy"—using food wastes (negative economic & environmental value) of the society to produce food (positive socio-economic value) for the society again" while also protecting farm soil and improving its fertility (positive economic & environmental value). And if vermicompost can "replace" the "chemical fertilizers" for production of "safe organic foods" which has now been proved worldwide, it will be a giant step towards achieving global social, economic & environmental sustainability. With the growing global popularity of "organic foods" which became a US $ 6.5 billion business every year by 2000, there will be great demand for vermicompost in future. US Department of Agriculture estimates 25% of Americans purchase organically grown foods at least once a week [42].

In all developed nations a "worm farm" has become a necessity in all wastewater & water treatment plants to resolve the problems of "sludge" which is a biohazard and needs safe disposal in secured landfills at high cost. Earthworms readily feed on them and convert into vermicompost.

In any vermiculture practice, "worm biomass" comes as a valuable by-product. It is finding uses and applications in modern medicine and in several industries for sustainable production of essential goods for societal use and consumption [107]. On commercial scale tons of worm biomass can result every year as under favorable conditions worms "double" their number at least every 60-70 days.

All infrastructure based on vermiculture technologies (vermicomposting, vermifiltration, vermiremediation & vermi-agroproduction) using earthworms are easy to construct, install and operate with minimum engineering considerations. They are highly economical and costeffective with highly valued by-products and end-products. It is basically a "one-time investment" technology as the earthworms multiply at a fast rate under favorable conditions of temperature and moisture and increase the pace and rapidity of the technological process.

Earthworms are truly justifying the beliefs and fulfilling the dreams of Sir Charles Darwin who called them as "unheralded soldiers' of mankind" and "friends of farmers" and said that "there may not be any other creature in world that has played so important a role in the history of life on earth". It is also justifying the beliefs of great Russian scientist Dr. Anatoly Igonin who said "Nobody and nothing can be compared with earthworms and their positive influence on the whole living Nature. They create soil & improve soil's fertility and provides critical biosphere's functions: disinfecting, neutralizing, protective and productive". Future of mankind on earth beholds with the earthworms and our relationship must be maintained.

ACKNOWLEDGEMENTS

The authors thank Mr. Scott Byrnes and Ms. Jane Gifkins—the in-charge of various laboratories and Mr. Ehsram Werner—the in-charge of glasshouses in the Faculty of Environmental Sciences of Griffith University, Nathan Campus for providing all laboratory and glasshouse facilities availed during the various research projects on vermiculture studies.

REFERENCES

1. M. Lotzof, "Vermiculture: An Australian Technology Success Story," Waste Management Magazine, Australia, February 2000.
2. R. K. Sinha, J. Nair, G. Bharambe, S. Patil and P. S. Bapat, "Vermiculture Revolution: A Low-Cost & Sustainable Technology for Management of Municipal & Industrial Organic Wastes (Solid & Liquid) by Earthworms with Significantly Low Greenhouse Gas Emissions," In: J. I. Daven and R. N. Klein, Eds., Progress in Waste Management Research, NOVA Science Publishers, Hauppauge, 2008, pp. 159-227.
3. R. K. Sinha, S. Herat, G. Bharambe, S. Patil, P. S. Bapat, K. Chauhan and D. Valani, "Vermiculture Biotechnology: The Emerging Cost-Effective and Sustainable Technology of the 21stCentury for Multiple Uses from Waste & Land Management to Safe & Sustained Food Production," Environmental Research Journal, NOVA Science Publishers, Hauppauge, Vol. 3, No. 1, 2009, pp. 41-110.
4. C. A. Edwards and J. R. Lofty, "Biology of Earthworms," Chapman & Hall, London, 1972, p. 283.
5. C. A. Edwards and P. J. Bohlen, "Biology and Ecology of Earthworms," 3rd Edition, Chapman and Hall, London, 1996.

6. C. Visvanathan, J. Trankler, K. Jospeh and R. Nagendran, (Eds.) "Vermicomposting as an Eco-Tool in Sustainable Solid Waste Management," Asian Institute of Technology, Annamalai University, Chidambaram, 2005.

7. P. Hand, "Earthworm Biotechnology," In: R. Greenshields, Ed., Resources and Application of Biotechnology: The New Wave, Macmillan Press Ltd, US, 1988.

8. J. E. Satchell, "Earthworm Ecology from Darwin to Vermiculture," Chapman and Hall Ltd., London, 1983, pp. 1-5.

9. M. Kerr and A. J. Stewart, "Tolerance Test of Eisinia fetida for Sodium Chloride," Journal of Undergraduate Research, U.S Department of Energy, 2002.

10. R. K. Sinha, "Earthworms: The Miracle of Nature (Charles Darwin's 'Unheralded Soldiers of Mankind and Farmer's Friends')," The Environmentalist, Vol. 29, No. 4, August 2009, pp. 339-340.

11. M. Appelhof, "Worms Eat My Garbage," 2nd Edition, Flower Press, Kalamazoo, Michigan, 1997. http://www. wormwoman. com

12. Z. W. Wang, "Research Advances in Earthworms Bioengineering Technology," Medica, Vol. 31, No. 5, 2000, pp. 386-389.

13. R. C. Loehr, J. H. Martin, E. F. Neuhauser and M. R. Malecki, "Waste Management Using Earthworms—Engineering and Scientific Relationships," Project Report ISP-8016764, National Science Foundation, Washington, D.C., 1984.

14. C. A. Edwards, "The Use of Earthworms in the Breakdown and Management of Organic Wastes," In: C. A. Edwards, Ed., Earthworm Ecology, CRC Press, Boca Raton, 1998, pp. 327-354.

15. M. T. Datar, M. N. Rao and S. Reddy, "Vermicomposting: A Technological Option for Solid Waste Management," Solid Waste Technology and Management, Vol. 24, No. 2, 1997, pp. 89-93.

16. G. Fraser-Quick, "Vermiculture—A Sustainable Total Waste Management Solution," What's New in Waste Management? Vol. 4, No. 6, 2002, pp. 13-16.

17. C. A. Edwards, "Breakdown of Animal, Vegetable and Industrial Organic Wastes by Earthworms," In: C. A. Edward and E. F. Neuhauser, Ed., Earthworms in Waste and Environmental Management, SPB Academic Publishing, The Hague, 1988, pp. 21-32.

18. R. D. Kale, "Earthworms: Nature's Gift for Utilization of Organic Wastes," In: C. A. Edward, Ed., Earthworm Ecology, St. Lucie Press, NY, 1998.

19. R. D. Kale and N. S. Sunitha, "Efficiency of Earthworms (E. Eugeniae) in Converting the Solid Waste from Aromatic Oil Extraction Industry into Vermicompost," Journal of IAEM, Vol. 22, No. 1, 1995, pp. 267-269.

20. S. N. Seenappa, J. Rao and R. Kale, "Conversion of Distillery Wastes into Organic Manure by Earthworm Eudrillus euginae," Journal of IAEM, Vol. 22, No. 1, 1995, pp.244-246.

21. K. Gunathilagraj and T. Ravignanam, "Vermicomposting of Sericultural Wastes," Madras Agricultural Journal, Coimbatore, 1996, pp. 455-457.

22. B. L. Lakshmi and G. S. Vizaylakshmi, "Vermicomposting of Sugar Factory Filter Pressmud Using African Earthworms Species (Eudrillus eugeniae)," Pollution Research, Vol. 19, No. 3, 2000, pp. 481-483.

23. M. Saxena, A. Chauhan and P. Asokan, "Flyash Vemicompost from Non-Friendly Organic Wastes," Pollution Research, Vol. 17, No. 1, 1998, pp. 5-11.

24. O. Bajsa, J. Nair, K. Mathew and G. E. Ho, "Pathogen Die-Off in Vermicomposting Process," Paper Presented at the International Conference on Small Water and Wastewater Treatment Systems, Perth, 2004.

25. R. K. Sinha, A. S. Herat, R. Asadi and E. Carretero, "Vermiculture Technology for Environmental Management: Study of Action of Earthworms Eisenia fetida, Eudrilus euginae and Perionyx excavatus on Biodegradation of Some Community Wastes in India and Australia," The Environmentalist, Vol. 22, No. 2, 2002, pp. 261-268.

26. R. K. Sinha, S. Herat, G. Bharambe and A. Brahambhatt, "Vermistabilization of Sewage Sludge (Biosolids) by Earthworms: Converting a Potential Biohazard Destined for Landfill Disposal into a Pathogen Free, Nutritive & Safe Bio-Fertilizer for Farms," Journal of Waste Management & Research, 2009.

27. R. K. Sinha, S. Herat, D. Valani, K. Singh and K. Chauhan, "Vermitechnology for Sustainable Solid Waste Management: A

Comparative Study of Vermicomposting of Food & Green Wastes with Conventional Composting Systems to Evaluate the Efficiency of Earthworms in Sustainable Waste Management with Reduction in Greenhouse Gas Emissions," NOVA Science Publications, Hauppauge, 2010.

28. M. C. Dash, "Role of Earthworms in the Decomposer System," In: J. S. Singh and B. Gopal, Eds., Glimpses of Ecology, India International Scientific Publication, New Delhi, 1978, pp. 399-406.

29. F. Binet, L. Fayolle and M. Pussard, "Significance of Earthworms in Stimulating Soilmicrobial Activity," Biology and Fertility of Soils, Vol. 27, No. 1, 1998, pp. 79-84.

30. D. R. Singleton, B. F. Hendrix, D. C. Coleman and W. B. Whitemann, "Identification of Uncultured Bacteria Tightly Associated with the Intestine of the Earthworms Lumricus rubellus," Soil Biology and Biochemistry, Vol. 35, 2003, pp. 1547-1555.

31. C. A. Edwards and K. E. Fletcher, "Interaction between Earthworms and Micro-Organisms in Organic Matter Breakdown," Agriculture Ecosystems and Environment, Vol. 24, No. 1-3, 1988, pp. 235-247.

32. M. Morgan and I. Burrows, "Earthworms/Microorganisms Interactions," Rothamsted Experimental Station, USA, 1982.

33. M. Appelhof, "Notable Bits," In: WormEzine, Vol. 2, No. 5, May 2003.

34. V. Pierre, R. Phillip, L. Margnerite and C. Pierrette, "Anti-Bacterial Activity of the Haemolytic System from the Earthworms Eisenia fetida andrei," Invertebrate Pathology, Vol. 40, No. 1, 1982, pp. 21-27.

35. R. Hartenstein and F. Hartenstein, "Physico-Chemical Changes Affected in Activated Sludge by the Earthworms Eisenia fetida," Journal of Environmental Quality, Vol. 10, No. 3, 1981, pp. 377-382.

36. R. K. Sinha, and A. Chan, "Study of Emission of Greenhouse Gases by Brisbane Households Practicing Different Methods of Composting of Food & Garden Wastes: Aerobic, Anaerobic and Vermicomposting," NRMA— Griffith University Project Report, 2009.

37. R. Sherman, "Commercial Systems Latest Development in Mid-to-Large Scale Vermicomposting," Biocycle, November 2000, p. 51.
38. G. Munroe, "Manual of on-Farm Vermicomposting and Vermiculture," Organic Agriculture Centre of Canada, 2007, p. 39.
39. P. Bogdanov, "Commercial Vermiculture: How to Build a Thriving Business in Redworms," VermiCo Press, Oregon, 1996, p. 83.
40. C. A. Edward, "Potential of Vermicomposting for Processing and Upgrading Organic Waste," Ohio State University, Ohio, 2000.
41. NCSU, "Large Scale Vermi-Composting Operations— Data from Vermi-Cycle Organics, Inc.," North Carolina State University, 1997.
42. K. M. Li and P. Z. Li, "Earthworms Helping Economy, Improving Ecology and Protecting Health," In: R. K. Sinha, et al., Eds., International Journal of Environmental Engineering (Special Issue on 'Vermiculture Technology'), 2010.
43. H. Mihara, M. Sumi, H. Mizumoto, T. Yoneta, R. Ikeda and M. Maruyama, "Oral Administration of Earthworm Powder as Possible Thrombolytic Therapy," Recent Advances in Thrombosis and Fibrinolysis, Academic Press, New York, 1990, pp. 287-298.
44. GEORG, "Feasibility of Developing the Organic and Transitional Farm Market for Processing Municipal and Farm Organic Wastes Using Large-Scale Vermicomposting," Good Earth Organic Resources Group, Halifax, Nova Scotia, 2004. http://www.alternativeorganic.com
45. J. Frederickson, "The Worm's Turn," Waste Management Magazine, UK, August 2000.
46. M. Gary, "Personal Communication from Envirofert (on Commercial Vermicomposting of Green Waste in New Zealand," 2009. http:www.envirofert.co.nz
47. S. Komarowski, "Vermiculture for Sewage and Water Treatment Sludge," WATER, July 2001.
48. U. S. Bhawalkar, "Vermiculture Eco-Technology," Publication of Bhawalkar Earthworm Research Institute (BERI), Pune, 1995.
49. P. K. Singh, "Production and Use of Vermicompost in India," College of Horticulture, Rajendra Agriculture University, Bihar, 2010.

50. R. D. Kale, "The Role of Earthworms and Research on Vermiculture in India," In: R. Guerrero and M. Guerrero, Eds., Vermitechnologies for Developing Countries, Proceedings of the International Symposium on Vermi Technologies for Developing Countries, Philippines, 2005, pp. 66-88.

51. R. Guerrero, "Commercial Vermimeal Production," In: R. Guerrero and M. Guerrero, Eds., Vermitechnologies for Developing Countries, Proceedings of the International Symposium on Vermi Technologies for Developing Countries, Philippines, 2005, p. 175.

52. S. Pajon, "The Worms Turn – Argentina," Intermediate Technology Development Group, Case Study Series 4, Munroe, 2007.

53. Z.-J. Sun, "Vermiculture and Vermi Protein," China Agricultural University Press, Beijing, 2003, p. 366.

54. A. Lopez and R. Alis, "Indigenous Use of Native Earthworms and its Fatty Acids Profile," Paper Presented at the Inernational Symposium on Vermitechnologies for Developing Countries, Laguna, Philippines, Also in Utilization of Earthworms for Health Remedies, 2005. http://www.wormsphilippines.com/docs/IKs%20on%20Earthworms.htm

55. I. N. Titov and B. M. Anokhin, "The Ten-Year Results of Treatment with the Extract of Earthworm Tissues," Innovation Centre, Moscow, Russia, In: R. Guerrero and M. Guerrero, Eds., Vermitechnologies for Developing Countries; Proceedings of the International Symposium on Vermi Technologies for Developing Countries, Philippines, 2005, pp. 148-149.

56. B. Tanaka and S. Nakata, "Studies of 'Antipyretic Components' from the Japanese Earthworm," Tokyo Igaku Zasshi, Vol. 29, 1974, pp. 67-97.

57. O. Bajsa, J. Nair, K. Mathew and G. E. Ho, "Vermiculture as a Tool for Domestic Wastewater Management," Water Science and Technology, IWA Publishing, Vol. 48, No. 11-12, 2003, pp. 125-132.

58. R. Hartenstein and M. S. Bisesi, "Use of Earthworm Biotechnology for the Management of Effluents from Intensively Housed Livestock," Outlook Agriculture, Vol. 18, No. 2, 1989, pp. 72-76.

59. M. Taylor, W. P. Clarke and P. F. Greenfield, "The Treatment of Domestic Wastewater Using Small-Scale Vermicompost Filter Beds," Ecological Engineering, Vol. 21, No. 2-3, 2003, pp. 197-203.
60. M. A. Soto and J. Toha, "Ecological Wastewater Treatment," Advanced Wastewater Treatment, Recycling and Reuse, AWT 98, Milano, 14-16 September 2008.
61. M. Xing, J. Yang and Z. Lu, "Microorganism-Earthworm Integrated Biological Treatment Process—A Sewage Treatment Option for Rural Settlements," ICID 21st European Regional Conference, Frankfurt, 15-19 May 2005.
62. R. K. Sinha, G. Bharambe and U. Chowdhary, "Sewage Treatment by Vermi-Filtration with Synchronous Treatment of Sludge by Earthworms: A Low-Cost Sustainable Technology over Conventional Systems with Potential for Decentralization," The Environmentalist, Springer, Vol. 28, No. 4, 8 April 2008, pp. 409-420.
63. R. K. Sinha, G. Bharambe and P. D. Bapat, "Removal of High BOD & COD Loadings of Primary Liquid Waste Products from Dairy Industry by Vermi-Filtration Technology Using Earthworms," Indian Journal of Environmental Protection, Vol. 27, No. 6, 2007, pp. 486-501.
64. V. Chandran and B. Soni, "Vermifiltration of Fruit Juice Processing Wastewater from in Brisbane," 40 CP Vermiculture Project, School of Engineering (Environment), Griffith University, Nathan Campus, Brisbane, Australia, (Supervisors Dr. Rajiv K. Sinha & Dr. Sunil Herat), 2010.
65. R. J. Hughes, J. Nair and K. Mathew, "The Implications of Wastewater Vermicomposting Technologies: On-Site Treatment Systems for Sustainable Sanitation," WAMDEC Conference, Zimbabwe, 27-30 July 2005.
66. S. I. Markman, A. Guschina, S. Barnsleya, L. Katherine, B. David and C.T. Muller, "Endocrine Disrupting Chemicals Accumulate in Earthworms Exposed to Sewage Effluents," Chemosphere, Vol. 70, No. 1, 2007, pp. 119- 125.
67. C. N. Lowe and K. R. Butt, "Inoculation of Earthworms into Reclaimed Soils: Experiences from Britain," Proceedings of 9th International Waste Management and Landfill Symposium, Sardinia, 2003.

68. K. R. Butt, C. N. Lowe, J. Frederickson and A. J. Moffat, "The Development of Sustainable Earthworm Populations at Calvert Landfill Site, UK," Land Degradation & Development, Vol. 15, No. 1, 2004, pp. 27-36.
69. R. Hartenstein, E. F. Neuhauser and J. Collier, "Accumulation of Heavy Metals in the Earthworm E. foetida," Environmental Quality, Vol. 9, No. 1, 1980, pp. 23-26.
70. W. C. Ma, J. Imerzeel and J. Bodt, "Earthworm and Food Interactions on Bioaccumulation and Disappearance of Pahs: Studies on Phenanthrene and Flouranthene," Ecotoxicology and Environmental Safety, Vol. 32, No. 3, 1995, pp. 226-232.
71. S. M. Contreras-Ramos, D. Alvarez-Bernal and L. Dendooven, "Eisenia fetida Increased Removal of Polycyclic Aromatic Hydrocarbons (PAHs) from Soil," Environmental Pollution, Vol. 141, No. 3, 2006, pp. 396-401.
72. Y. Tomoko, K. Toyota and S. Hiroaki, "Enhanced Bioremediationof Oil-Contaminated Soil by a Combination of the Earthworm (Eisenia Fetida) and Tea Extraction Residue," Edaphologia, Vol. 77, 2005, pp. 1-9.
73. M. Schaefer, "Earthworms in Crude Oil Contaminated Soils: Toxicity Tests and Effects on Crude Oil Degradation," Contaminated Soil Sediment & Water, Vol. 35, 2005, pp. 7-8.
74. J. Martin-Gil, L. M. Navas-Gracia, E. Gomez-Sobrino, A. Correa-Guimaraes, S. Hernandez-Navarro and M. Sanchez-Bascones, "Composting and Vermicomposting Experiences in the Treatments and Bioconversion of Asphaltens from the Prestige Oil Spill," Journal of Bioresource Technology, Vol. 99, No. 6, 2007, pp. 1821-1829.
75. B. Davis, "Laboratory Studies on the Uptake of Dieldrin and DDT by Earthworms," Soil Biology and Biochemistry, Vol. 3, 1971, pp. 221-223.
76. M. P. Ireland, "Heavy Metals Uptake in Earthworms," Earthworm Ecology, Chapman & Hall, London, 1983.
77. J. Haimi, J. Salminen, V. Huhta, J. Knuutinen and H. Palm, "Bioaccumulation of Organochlorine Compounds in Earthworms," Soil Biology & Biochemistry, Vol. 24, No. 12, 1992, pp. 1699-1703.

78. N. S. Bolan and S. Baskaran, "Characteristics of Earthworm Casts Affecting Herbicide Sorption and Movement," Biological Fertility of Soils, Vol. 22, No. 4, 1996, pp. 367-372.

79. B. Gevao, C. Mordaunt, K. T. Semple, T. G. Piearce and K. C. Jones, "Bioavailability of Nonextractable (Bound) Pesticide Residues to Earthworms," Environmental Science & Technology, Vol. 35, No. 3, 2001, pp. 501-507.

80. A. C. Singer, W. Jury, E. Leupromchai, C.-S. Yahng and D. E. Crowley, "Contribution of Earthworms to PCB Bioremediation," Journal of Soil Biology & Biochemistry, Vol. 33, No. 6, 2001, pp. 765-775.

81. R. K. Sinha, G. Bharambe and D. Ryan, "Converting Wasteland into Wonderland by Earthworms: A Low-Cost Nature's Technology for Soil Remediation: A Case Study of Vermiremediation of PAH Contaminated Soil," The Environmentalist, Vol. 28, No. 4, 14 May 2008, pp. 466-475.

82. K. R. Butt, "Inoculation of Earthworms into Reclaimed Soils: The UK Experience," Land Degradation and Development, Vol. 10, No. 6, 1999, pp. 565-575.

83. G. B. Brown and B. M. Doube, "On Earthworms Assisted Bioremediation," In: C. A. Edward, Ed., Earthworm Ecology, 2nd Edition, CRC Press, Boca Raton, 2004, pp. 213-239.

84. R. D. Kale and K. Bano, "Field Trials with Vermicompost: An Organic Fertilizer," Proceedings of National Seminar on Organic Waste Utilization by Vermicomposting, GKVK Agricultural University, Bangalore, 1986.

85. C. A. Edwards and I. Burrows, "The Potential of Earthworms Composts as Plant Growth Media," In: C. A. Edward and E. F. Neuhauser, Eds., Earthworms in Waste and Environmental Management, SPB Academic Publishing, The Hague, The Netherlands, 1988, pp. 21-32.

86. L. P. Canellas, F. L. Olivares, A. L. Okorokova and R. A. Facanha, "Humic Acids Isolated from Earthworm Compost Enhance Root Elongation, Lateral Root Emergence, and Plasma Membrane H^+-Atpase Activity in Maize Roots," Plant Physiology, Vol. 130, No. 4, 2002, pp. 1951-1957.

87. H. I. Chaoui, L. M. Zibilske and T. Ohno, "Effects of Earthworms Casts and Compost on Soil Microbial Activity and Plant Nutrient Availability," Soil Biology and Biochemistry, Vol. 35, No. 2, 2003, pp. 295-302.

88. G. H. Baker, P. M. Williams, P. J. Carter and N. R. Long, "Influence of Lumbricid Earthworms on Yield and Quality of Wheat and Clover in Glasshouse Trials," Journal of Soil Biology and Biochemistry, Vol. 29, No. 3-4, 1997, pp. 599-602.

89. S. Palaniswamy, "Earthworm and Plant Interactions," Paper Presented in ICAR Training Program, Tamil Nadu Agricultural University, Coimbatore, 1996.

90. N. Q. Arancon, C. A. Edwards, P. Bierman, C. Welch and J. D. Metzger, "Influences of Vermicomposts on Field Strawberries-1: Effects on Growth and Yields," Bioresource Technology, Vol. 93, No. 2, 2004, pp. 145-153.

91. K. A. Webster, "Vermicompost Increases Yield of Cherries for Three Years after a Single Application," EcoResearch, South Australia, 2005.

92. J. C. Buckerfield and K. A. Webster, "Worm-Worked Waste Boost Grape Yield: Prospects for Vermicompost Use in Vineyards," The Australian and New Zealand Wine Industry Journal, Vol. 13, No. 1, 1998, pp. 73-76.

93. R. K. Sinha, S. Herat, D. Valani and K. Chauhan, "Vermiculture and Sustainable Agriculture," American-Eurasian Journal of Agricultural and Environmental Sciences, IDOSI Publication (Special Issue), 2009, pp. 1-55.

94. A. A. Ansari, "Effect of Vermicompost on the Productivity of Potato (Solanum tuberosum) Spinach (Spinacia oleracea) and Turnip (Brassica campestris)," World Journal of Agricultural Sciences, Vol. 4, No. 3, 2008, pp. 333- 336.

95. C. A. Edwards and N. Arancon, "Vermicompost Suppress Plant Pests and Disease Attacks," REDNOVA NEWS, 2004.

96. G. F. de Boer and O. Sova, "Vermicomposting as a Resource for Biodegradable Detergents," 4th ZERI World Congress, Windhoek, Namibia, 1998.

97. C. H. Cordero, "Earthworms Can Help Dissolve Blood Clots for Stroke Patients," 2005.

98. W. L. Cheng and Z. J. Sun, "Pharmaceutical Value and Uses of Earthworms," Vermillenium Abstracts, Flowerfield Enterprizes, Kalamazoo, 2000.
99. [E. Cooper, "New Enzymes Isolated from Earthworms is Potent Fibrinolytic," ACAM Integrative Medicine Blog, Oxford University Press Journal, 2009.
100. C. M. Hwang, D. Kim and S. H. Huh, "In-vivo Evaluation of Lumbrokinase Extracted from Earthworms Lumbricus rubellus in a Prosthetic Vascular Graft," Cardiovascular Surgery, Vol. 43, No. 6, 2002, pp. 891-894.
101. C. Qingsui, "A New Medicine for Heart Diseases Containing Enzyme Activator Extracted from Earthworms," In: Lopez & Alis, The Utilization of Earthworms for Health Remedies, 2003.
102. H. Mihara, M. Maruyama and H. Sumi, "Novel Thrombolytic Therapy Discovered in Oriental Medicine Using the Earthworms," SE Asian Journal of Tropical Medicine & Health, Vol. 23, Suppl 2, 1992, pp. 131-140.
103. S. L. Li, "Research on di long's (Earthworms) Effect in Lowering Blood Pressure," Journal of Information, Vol. 12, No. 3, 1995, pp. 22-24.
104. R. Moss, "Of Enzymes, Worms & Cancer: The War on Cancer (Lumbrokinsae enzyme from Earthworms)," Worm Digest, 2004.
105. L. Jin, H. Jin, G. Zhang and G. Xu, "Changes in Coagulation and Tissue Plasminogen Activator after the Treatment of Cerebral Infarction with Lumbrokinsae (from Earthworms)," Clinical Hemorheology and Microcirculation, Vol. 23, 2000, pp. 213-218.
106. UNSW, ROU, "Best Practice Guidelines to Managing On-Site Vermiculture Technologies," University of New South Wales Recycling Organics Unit, Sydney, 2002.
107. R. A. Dynes, "Earthworms: Technology Information to Enable the Development of Earthworm Production," Rural Industries Research and Development Corporation, Government of Australia, Canberra, 2003.

Chapter 5

Optimization and Comparative Study of the Sugar Waste for the Growth of Rhizobium Cells Along with Traditional Laboratory Media

[1]Ashok Kumar Singh,[1]Gauri,[2]Rajendra Prasad Bhatt, and[1]Shailja Pant

[1]Department of Microbiology, Dolphin (P.G) Institute of Biomedical and Natural Sciences, Manduwala, Dehradun (Uttarakhand), India

[2]Department of Botany and Microbiology, H.N.B. Garhwal University (A Central University), SriNagar, Garhwal (Uttarakhand), India

ABSTRACT

To produce good quality nitrogen inoculants, it is desired to obtain effective rhizobia with good nodulation and nitrogen fixing capacity to the host plant. It is therefore, required to find a nutrient medium for their growth and optimize the pH, temperature, incubation period and aeration. This study was conducted to evaluate the use of sugar waste (molasses) as an alternative growth medium for the cultivation of *Rhizobium* and to determine the optimum environmental parameters such as pH and temperature. Growth and population count of *Rhizobium trifolii* (MTCC 905) at different concentrations of sugar waste was monitored by recording Optical Density (OD) and Colony forming unit (cfu mL^{-1}). Growth and population count were highest at 10% concentration i.e., 0.706±0.012 and 8.98±0.12, respectively. After optimizing the concentration of sugar waste growth pattern of *R. trifolii* was observed at 10% sugar waste concentration along with different synthetic media by recording Optical Density (OD) at 600 nm after 12 to 60 h. Cells of fast growing rhizobia grow rapidly on medium containing only 10% sugar waste and growth was superior to that of the control (standard media) used for *Rhizobium*. A pH of 7.0, temperature of 28°C and 48 h incubation period were most appropriate for *Rhizobium trifolii* (MTCC 905). The present study concludes that 10% sugar waste is complete medium for the rhizobial inoculants.

INTRODUCTION

The economic and the environmental costs of agricultural production have been increased by non-judicious use of chemical fertilizers and its techniques of production and application. Consequently, the environmental problems have magnified due to the pollution of air, soil and water sources. The biological fertilizers may be incorporated as additional tools which can be economical and clean alternatives for the sustainable management of ecosystems (Mishra *et al.*, 2006). Thus, the initial step in the legume inoculant production is the large number of selected *Rhizobium* sp. in liquid medium (Thompson, 1991). Therefore, it is the need of the present scenario to develop an economical as well as productive formulation and nutrient medium for the growth of such rhizobial cells. Most of the researchers choose YEM

(Yeast Extract Mannitol) medium for rhizobial culture which contains mannitol or glycerol as a carbon source, yeast extract as a source of nitrogen, mineral salts and growth factors (Verma *et al.*, 2010; Fahmi *et al.*, 2011; Yadav *et al.*, 2011) due to these costly ingredients of that medium making it not suitable for commercial production of biofertilizers (Rebah *et al.*, 2002).

In view of the growing demand of rhizobial inoculants it is imperative to search cheap and readily available substances to these costly ingredients. Numerous investigators have looked for ways of producing biofertilizers using low cost media. A variety of agricultural and industrial by-products such as proteolyzed pea husks, molasses and water hyacinth (Gulati, 1979), malt sprouts (Bioardi and Ertola, 1985), deproteinized leave extracts (Chanda *et al.*, 1987), cheese whey (Estrella *et al.*, 2004), waste water sludge (Ben Rebah *et al.*, 2007) and Jaggery solution (Jain *et al.*, 2000). These products contain growth factors, carbon and/or nitrogen source and to support the growth of rhizobia equal to or better than the known growth in the available media. Molasses is also used as a substrate for the production of Biosurfactants byPanesar *et al.* (2011).

Molasses is the final effluent obtained in the preparation of sucrose by repeated evaporation, crystallization and centrifugation of juices from sugar cane and sugar beat (Tunuguntla and Sullivan, 2004). Beside carbon and nitrogen source molasses is also an excellent source of manganese, copper, iron, calcium, potassium, magnesium, vitamin B6 and selenium (Aslan *et al.*, 1997) and have high Ash and crude proteincontent (Paviz *et al.*, 2011). The objective of the present study was to optimize the molasses as a cultivation medium and growth conditions (pH and temperature) for *Rhizobium trifolii* (MTCC 905).

MATERIALS AND METHODS

This study was conducted in Microbiology Department of Dolphin (P.G) Institute of Biomedical and Natural Science, Dehradun (Uttarakhand), India in the year 2009.

Rhizobium trifolii (MTCC 905) was used as model strain and obtained from Microbial Type Culture Collection (MTCC) from the Institute of Microbial Technology (IMTECH), Chandigarh, India.

Sugar Waste as a Substrate

The sugar waste was collected from Uttarakhand State Sugar Corporation Ltd. Unit: Doiwala, Dehradun, Uttrakhand.

Chemical Analysis of Sugar Waste

The pH, Dissolved Oxygen (DO), Biochemical Oxygen Demand (BOD), Chemical Oxygen Demand (COD) and reducing sugar content of sugar waste was studied (APHA, 1992).

Starter Culture

The culture (50 mL) was prepared by inoculating *Rhizobium* strain into Tryptone Yeast extract (TY) broth and *Rhizobium* Minimal Medium (RMM), respectively and incubated at 28±2°C for 16-24 h.

Media Optimization

Optimization of Sugar Waste

The starter culture (1%) was inoculated into different concentration of sugar waste i.e., 10, 20 up to 100% and growth was monitored by recording Optical Density (OD) and colony forming unit (cfu mL^{-1}) at regular time interval. The viable count of bacterial population was also observed (Cappuccino and Sherman, 2001).

Comparison of Sugar Waste with Different Lab Media

Effect of different media i.e., 10% sugar waste, RMM and TY on *Rhizobium* was monitored in terms of absorbance at 600 nm after every 12 up to 60 h by inoculating the bacteria with these media.

Effect of Culture Parameters on Growth

pH

Rhizobial culture was inoculated into 10% sugar waste media having constant temperature maintained at different pH i.e., 6.0, 6.5, 7.0, 7.5, 8.0. The growth was monitored in terms of absorbance at 600 nm after every 12 up to 60 h. The viable count of bacterial population was also observed.

Temperature

At constant pH of 7.0 and 10% sugar waste concentration, *Rhizobium* strain was inoculated into 10% sugar waste as a suitable growth media maintained at different temperature i.e., 26, 27, 28, 29, 30°C and growth was observed by recording Optical Density (OD) at 600 nm after every 12 up to 60 h. The viable count of bacterial population was also observed. The data were expressed in Mean±SE of mean values.

RESULTS

Physiochemical Characteristics of Sugar Waste

Three representative samples of sugar waste were tested for physiochemical characteristics. The sugar waste was brown in color with a pH of 6.8±0.2 and Dissolved oxygen 40 mg L^{-1}. The Biochemical Oxygen Demand (BOD) and Chemical Oxygen Demand (COD) were found 232 and 450 mg L^{-1}, respectively. The reducing sugar content of sugar waste was 2.75±0.17 g L^{-1}.

Effect of Concentration of Sugar Waste on Growth and Population Count of Rhizobium Trifolii (Mtcc 905)

Keeping the two parameters constant i.e., pH and Temperature, 7.0 and 28°C, respectively. Growth and population count of *Rhizobium trifolii* (MTCC 905) at different concentrations of sugar waste (10, 20, 30, 40, 50, 60, 70, 80, 90 and 100%) was monitored by recording Optical Density (OD) at 600 nm and after 48 h incubation. There was a considerable decline in OD values and cfu mL^{-1} after 10% sugar waste concentration. At 10% concentration OD values and cfu mL^{-1} are 0.706±0.012 and 8.98±0.12, respectively. However, effect of sugar waste concentration (above 10%) was clearly visible on the growth and population count of bacteria (Fig. 1 and 2). Minimum growth for *Rhizobium trifolii* was observed at 90% sugar waste concentration i.e., 0.004±0 in terms of OD and at 70% sugar waste concentration i.e., 0.21±0.05 in terms of Colony forming unit mL^{-1}.

Figure 1 represents the growth of bacteria in terms of optical density while Fig. 2 represents the population count of bacteria in terms of Colony forming unit mL^{-1}.

Growth Profile of Rhizobium Trifolii *(MTCC 905) on Sugar Waste Concentrations and Synthetic Media*

Keeping the two parameters constant i.e., pH and Temperature, 7.0 and 28°C, respectively. Growth of *Rhizobium trifolii* was observed at 10% sugar waste concentration along with RMM and TY Medium separately within same conditions by recording Optical Density (OD) at 600 nm after every 12 h

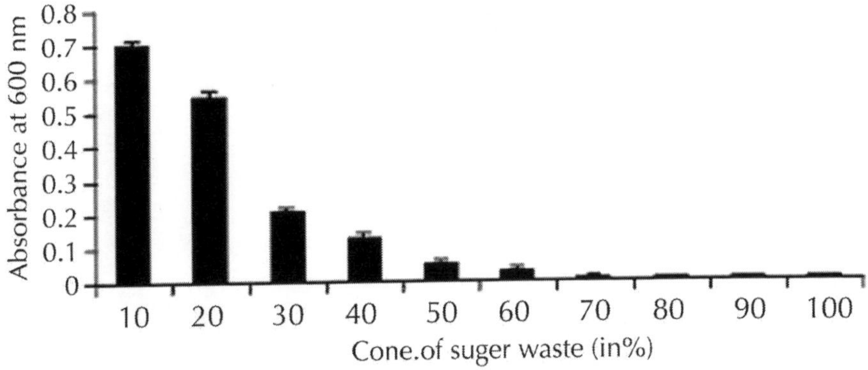

Figure 1: Effect of concentration of sugar waste on growth of Rhizobium trifolii (48 h incubation)

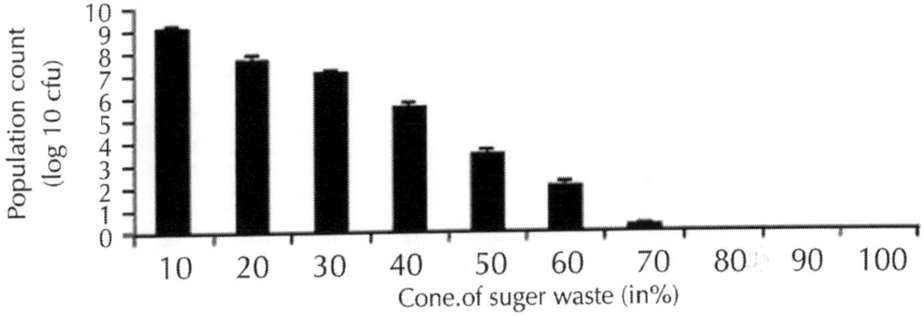

Figure 2: Effect of concentration of sugar waste on population of Rhizobium trifolii

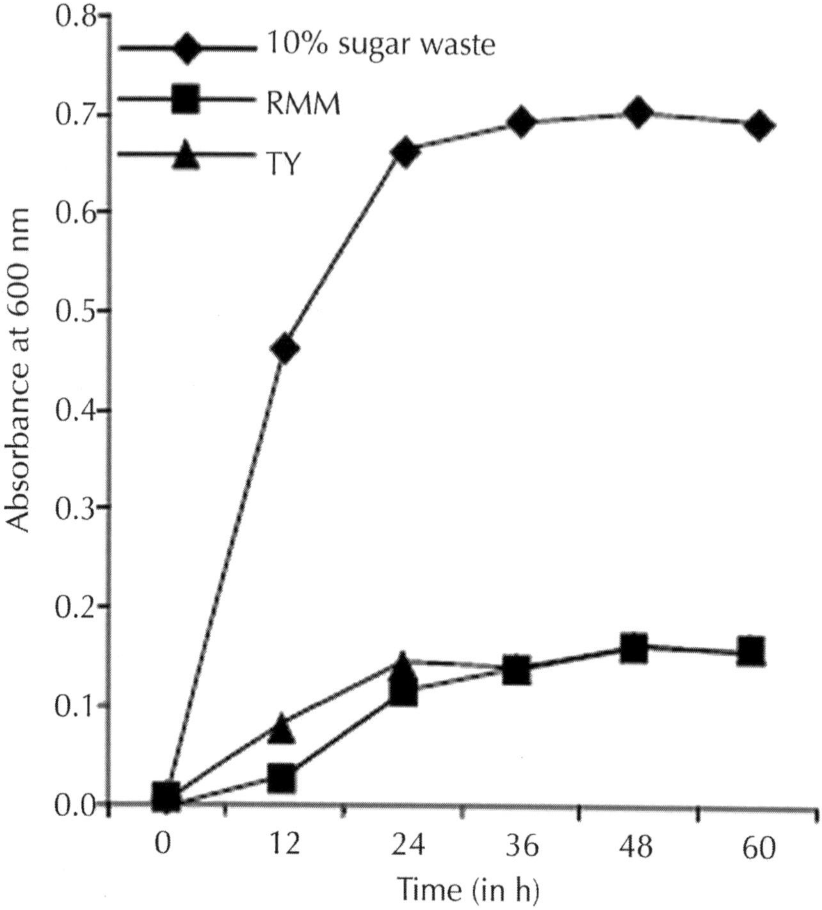

Figure 3: Growth profile of Rhizobium trifolii on sugar waste and synthetic medium

Growth of Rhizobium trifolii was found maximum in 10% sugar waste concentration at 48 h i.e., 0.706±0.011 and minimum in RMM Medium i.e., 0.162±0.014 (Fig. 3). The number of bacteria reached maximum at 48 h and above which the growth was constant.

Growth and Population Count of **Rhizobium Trifolii** ***(Mtcc 905) at Different pH***

Keeping the two parameters constant i.e., concentration of sugar waste and Temperature, 10% and 28°C, respectively. The growth of

Rhizobium trifolii at different pH (i.e., 6.0, 6.5, 7.0, 7.5 and 8.0) was recorded in terms of optical density (600 nm) after every 12 up to 60 h (Fig. 4). There was a considerable increase in OD values reaching a maximum of 0.706±0.009 with increasing pH up to 7.0 in 48 h, above which the growth was constant. However, inhibitory effect of acidic pH (below 7.0) and alkaline pH (above 7.0) was clearly visible on the growth. Minimum growth was observed at pH 6.0 and 8.0 i.e., 0.512±0.015. Population count of Rhizobium trifolii at different pH was monitored in terms of cfu mL^{-1} after 48 h incubation. There was a considerable increase in cfu values reaching a maximum of 8.48±0.003 with increasing pH up to 7.0 after 48 h (Fig. 5). Minimum growth was observed at pH 6.0 i.e., 8.133±0.006 in terms of cfu mL^{-1}.

Effect of Different Temperature on Growth and Population Count of Rhizobium Trifolii *(MTCC 905)*

When the two parameters was kept constant i.e., pH and concentration of sugar waste, 7.0 and 10%, respectively, growth of Rhizobium trifolii at different temperature (i.e., 26, 27, 28, 29 and 30°C) was monitored by recording Optical Density (OD) at 600 nm after every 12 up to 60 h (Fig. 6).

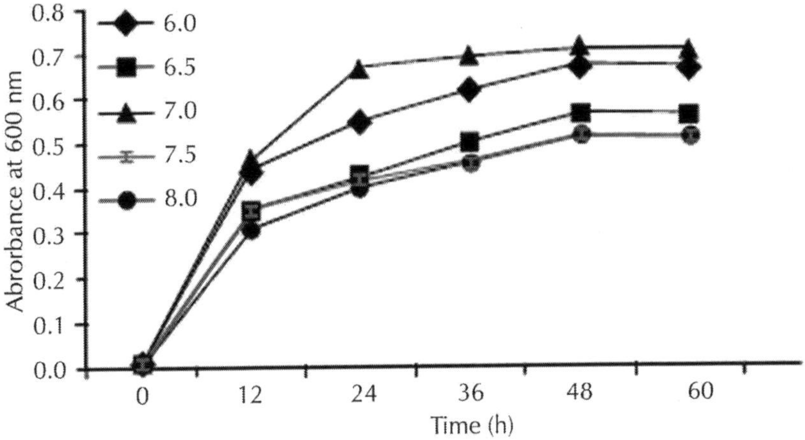

Figure 4: pH dependent growth profile of Rhizobium trifolii (Temperature 28°C, 10% sugar waste)

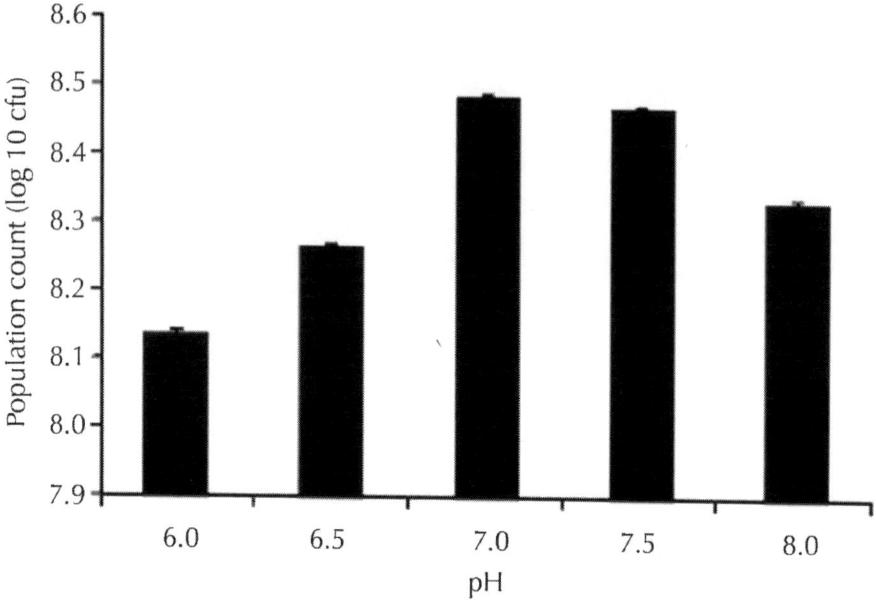

Figure.5: Effect of pH on population of Rhizobium trifolii (Temperature -28°C, sugar waste-10%)

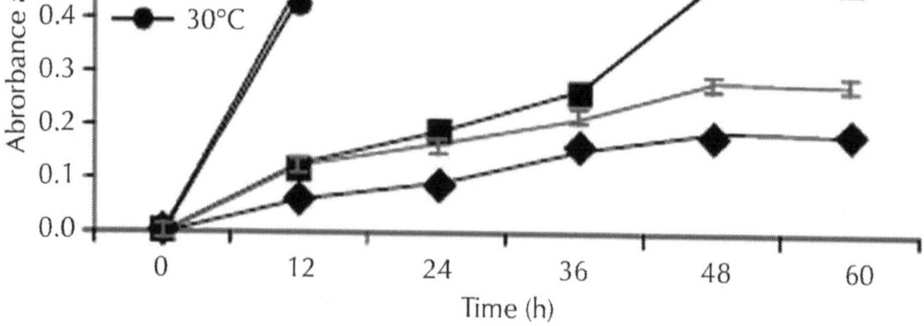

Figure 6: Temperature dependent growth profile of Rhizobium trifolii (pH 7.0, 10% sugar waste)

There was a significant increase in OD values reaching a maximum of 0.706±0.012 with the increase in temperature up to 28°C in 48 h. However, effect of temperature (below and above 28°C) was clearly visible on the growth. Minimum growth of Rhizobium trifolii was observed at 26°C i.e., 0.187±0.012. Growth of Rhizobium trifolii (MTCC 905) after 48 h was constant. Population count of Rhizobium trifolii at different temperature was monitored in terms of cfu mL^{-1} after 48 h incubation. There was a considerable increase in cfu values reaching a maximum of 8.98±0.12 with increasing temperature up to 28°C after 48 h (Fig. 7)

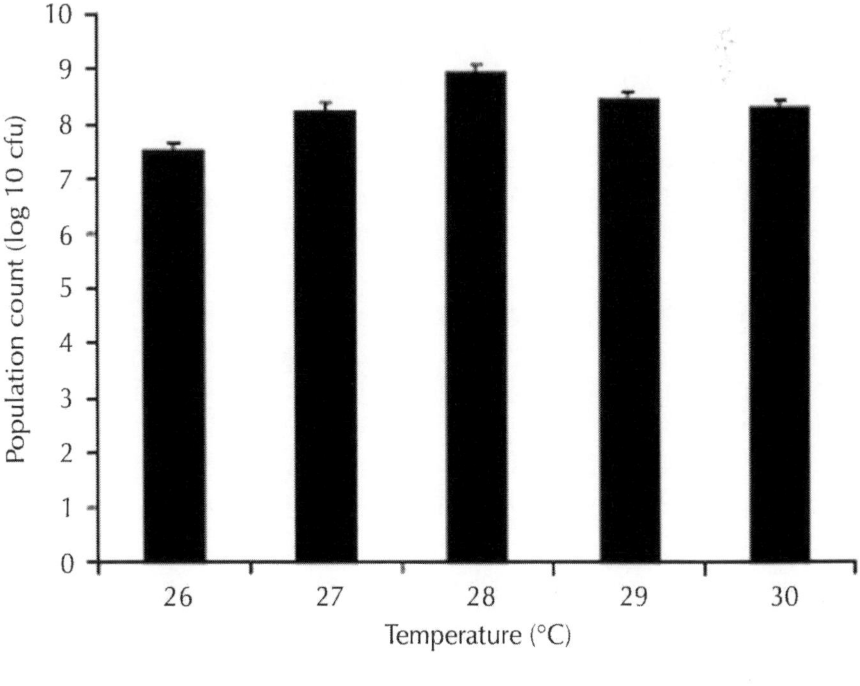

Figure 7: Effect of temperature on population of Rhizobium trifolii (pH 7.0, sugar waste 10%)

DISCUSSION

The sugar waste obtained from Uttarakhand State Sugar Corporation Ltd. Unit. Doiwala, Dehradun was analyzed for various physiochemical parameters such as pH, Dissolved Oxygen (DO), Biochemical Oxygen

Demand (BOD), Chemical Oxygen Demand (COD) and reducing sugar content. Simurina et al. (2008) also analyzed physiochemcial parameters of sugarwaste. Growth and population count of R. trifolii was observed at different concentrations of sugar waste (10, 20, up to 100%) by recording Optical Density (OD) at 600 nm and log 10 (cfu mL^{-1}), respectively. Our results showed that molasses could serve as a better substrate for growth of bacteria in terms of both OD and cfu Nandi and Sinha (1974) conducted the experiments for mass scale production of Rhizobium inoculants in different cheap carbon sources and in different concentrations of yeast extracts in the medium. Their findings suggest molasses could serve as a superior media for Rhizobium as molasses contain organic substances and trace elements which helps the growth of bacteria. It has been observed in our results that concentration of sugar waste is indirectly proportional to the Optical density as well as cfu. According to Baei et al. (2009) molasses are obtained from circulation of sugar solution in series of evaporation, it contains caramelized and invert sugars, if high sugar waste concentration (as a substrate) is implemented, it may cause cell toxicities.

At above 10% sugar waste concentration, fall in growth and population count of Rhizobium was observed which might be due to that at higher concentrations of sugar waste and improper aeration conditions in the medium. The aerobic condition is good for nitrogen fixing bacteria (Hassen et al., 2007). Therefore, sugar waste concentration of 10% was taken optimum for the study.

After optimizing the concentration of sugar waste i.e., 10% growth of R. trifolii was observed at 10% sugar waste concentration along with different standard media by recording Optical Density (OD) at 600 nm after 12 to 60 h. According to present study growth of R. trifolii were rapidly increased after 12 h, thereafter, the growth was slowly increased, reaching maximum at 48 h and then attained stationary phase. The results of this study indicates that on medium containing only 10% sugar waste, cells of fast growing rhizobia grow rapidly and growth was superior to that of the control (standard media). These results demonstrates that 10% sugar waste can be satisfactorily used as a growth substrate of rhizobia. Estrella et al. (2004) found that cheese whey based growth medium efficiently protects Rhizobium loti cells during freezing and can maintain their growth in a manner similar to that of traditional mannitol-based medium (YEM). Ben Rebah et al.

(2007) reported that waste water sludge, a world wide recyclable waste showed good potential for inoculants production as a growth medium.

pH is an important parameter for the growth of organisms, slight variation in pH of medium might have enormous effects on the growth of organisms, keeping this in mind, at different pH (6.0, 6.5, 7.0, 7.5 and 8.0) growth and population count of fast growing strain of R. trifolii was monitored at constant temperature (28°C) and sugar waste concentration (10%). The result of the effects of different pH values on the growth and population count of R. trifolii in culture shows that these strains grow well at pH between 6.0-8.0 with maximum at pH 7.0 on observing OD and c.f.u. This is in accordance with Mensah et al. (2006). They recorded maximum absorbance for broth experiment as well as exhibited heavy growth as measured by population count at pH 7.0 for Rhizobium species. Ali et al. (2009) also observed that there was considerable increase in OD values with increasing pH up to 7.0.2. This is according to the earlier findings (Hussain et al., 2002; Yadegari et al., 2008).

Temperature also plays an important role in the growth of an organism, so after optimization of pH (7.0) and sugar waste concentration (10%), growth of R. trifolii was observed at different temperatures (26, 27, 28, 29, 30°C). In our results it was observed that fast growing strains grow well at temperature between 26-30°C, when growth was measured in terms of OD and c.f.u. after 48 h incubation period. It has been reported in literature that temperature for growth of root nodulating bacteria ranged from 25-0°C (Pongsilp et al., 2010). In our study maximum growth of R. trifolii was observed at 28°C therefore, it is assumed to be the most suitable temperature for Rhizobium. This is in agreement with the earlier findings (Ogutco et al., 2008; Agarwal and Ahmad, 2010; Parthiban et al., 2011).

CONCLUSIONS

The present study recommends that the use of 10% sugar waste (molasses) as a suitable growth media because it is readily available, inert, cheap, ecofriendly and superior to the standard media used for Rhizobium. It use as an alternate of synthetic media in commercial preparations of biofertilizers will considerably reduce the production cost.

REFERENCES

1. APHA, 1992. Standard Methods for the Examination of Water and Wastewater. 18th Edn., APHA, WEF and AWWA, Washington DC., USA.
2. Agarwal, S. and Z. Ahmad, 2010. Contribution of the Rhizobium inoculation on plant growth and productivity of two cultivars of berseem (Trifolium alexandrinum L.) in saline soil. Asian J. Plant Sci., 9: 344-350.
3. Ali, S.F., L.S. Rawat, M.K. Meghvansi and S.K. Mahna, 2009. Selection of stress-tolerant rhizobial isolates of wild legumes growing in dry regions of Rajasthan, India. ARPN J. Agric. Biol. Sci., 4: 13-18.
4. Aslan, Y., E. Erduran, H. Mocan, Y. Gedik, A. Okten, H. Soylu and O. Deger, 1997. Absorption of iron from grape molasses and ferrous sulfate: A comparative study in normal subjects and subjects with iron deficiency anemia. Turk. J. Pediatr., 39: 465-471.
5. Baei, M.S., G.D. Najafpour, H. Younesi, F. Tabandesh and H. Eisazadeh, 2009. Poly (3-hydroxybutyrate) synthesis by Cupriavidus necator DSMZ 545 utilizing various carbon sources. World Applied Sci. J., 7: 157-161.
6. Ben Rebah, F., D. Prevost, A. Yezza and R.D. Tyagi, 2007. Agro-industrial waste materials and waste water sludge for rhizobial inoculant production: A review. Bioresour. Technol., 98: 3535-3546.
7. Bioardi, J.L. and R.J. Ertola, 1985. Rhizobium biomass production in batch and continous culture with a malt-sprouts medium. World J. Microbiol. Biotechnol., 1: 163-172.
8. Cappuccino, S. and N. Sherman, 2001. Serial Dilution Agar Plating Procedure to Quantitate Viable Cells. In: Microbiology: A Laboratory Manual, Cappuccino, J.G. and N. Sherman (Eds.). 6th Edn., Benjamin-Cummings Publishing Co., San Francisco, CA., pp: 119-124.
9. Chanda, S., S. Matai and S. Chakrabatri, 1987. Deproteinized leaf juice as a medium for growth of Rhizobium. Indian J. Exp. Biol., 25: 573-575.

10. Estrella, M.J., F.L. Pieckenstain, M. Marina, L.E. Diaz and O.A. Ruiz, 2004. Cheese whey: An alternative growth and protective medium for Rhizobium loti cells. J. Indian Microbiol. Biotechnol., 31: 122-126.
11. Fahmi, A.I., H.H. Nagaty, R.A. Eissa and M.M. Hassan, 2011. Effects of salt stress on some nitrogen fixation parameters in faba bean. Pak. J. Biol. Sci., 14: 385-391.
12. Gulati, S.L., 1979. New nonsynthetic medium for Rhizobium culture production from wastes. Biotechnol. Bioeng., 21: 1507-1515.
13. Hassen, A.A., J. Xu and J. Yang, 2007. Growth conditions of associative nitrogen-fixing bacteria Enterobacter cloace in rice plants. Agric. J., 2: 672-675.
14. Hussain, N., F. Mujeeb, M. Tahir, G.D. Khan, N.M. Hassan and A. Bari, 2002. Effectiveness of rhizobium under salinity stress. Asian J. Plant Sci., 1: 12-14.
15. Jain, S.K., D.V. Pathak and H.R. Sharma, 2000. Alternate carbon substrate for mass production of Rhizobium inoculants. Haryana Agric. Univ. J. Res., 30: 1-6.
16. Mensah, J.K., F. Esumeh, M. Iyamu and C. Omoifo, 2006. Effects of different salt concentrations and pH on growth of Rhizobium sp. and a cowpea `Rhizobium association. Am.-Eurasian J. Agric. Environ. Sci., 3: 198-202.
17. Mishra, R.P., R.K. Singh, H.K. Jaiswal, V. Kumar and S. Maurya, 2006. Rhizobium-Mediated induction of phenolic and plant growth promotion in rice (Orizza Sativa L). Curr. Microbiol., 52: 383-389.
18. Nandi, P.N. and N. Sinha, 1974. Mass scale production of Rhizobium culture. Proc. Indian Natl. Acad., 40: 479-481.
19. Ogutco, H., O.F. Algur, E. Elkoca and F. Kantar, 2008. The determination of symbiotic effectiveness of Rhizobium strains isolated from Wild Chickpeas collected from high altitudes in Erzurum. Turk. J. Agric. For., 32: 241-248.
20. Panesar, R., P.S. Panesar and M.B. Bera, 2011. Development of low cost medium for the production of biosurfactants. Asian J. Biotechnol., 3: 388-396.

21. Parthiban, K., S. Manikandan and S. Ganesapandian, 2011. Production of cellulose I microfibrils from Rhizobium sp. and its wound healing activity on mice. Asian J. Applied Sci., 4: 247-254.

22. Paviz, M.M., T. Ghoorchi and F. Ghanbari, 2011. Effects of molasses and bacterial inoculant on chemical composition and aerobic stability of sorghum silage. Asian J. Anim. Vet. Adv., 6: 385-390.

23. Pongsilp, N., C. Leelahawonge, A. Nuntagij, N. Teaumroong and N. Boonkerd, 2010. Characterization of Pueraria mirifica nodulating rhizobia present in Thai soil. Afr. J. Microbiol. Res., 4: 1307-1313.

24. Rebah, F.B., D.T. Rajeshwar, P. Danielle and Y.S. Rao, 2002. Wastewater sludge as a new medium for rhizobial growth. Water Qual. Res. J. Canada, 372: 353-370.

25. Simurina, O.D., B.V. Filipev, L.B. Levi and V.D. Pribis, 2008. Application of sugar beet molasses in the production of tea biscuits. Food Process. Qual. Saf., 4: 201-206.

26. Thompson, J.A., 1991. Legume Inoculant Production and Quality Control. In: Report on the Expert Consultation on Legume Inoculant Production and Quality Control, Thompson, J.A. (Ed.). FAO, Rome, pp: 15-32.

27. Tunuguntla, A. and M.J. Sullivan, 2004. Black strap molasses for the treatment of inflammatory bowel disease-associated anemia. South Med. J., 97: 794-794.

28. Verma, J.P., J. Yadav and K.N. Tiwari, 2010. Application of Rhizobium sp. BHURC01 and plant growth promoting rhizobactria on nodulation, plant biomass and yields of Chickpea (Cicer arietinum L.). Int. J. Agric. Res., 5: 148-156.

29. Yadav, J., J.P. Verma, V.K. Rajak and K.N. Tiwari, 2011. Selection of effective indigenous Rhizobium strain for seed inoculation of chickpea (Cicer aritenium L.) production. Bacteriol. J., 1: 24-30.

30. Yadegari, M., H.A. Rahmani, G. Noormohammadi and A. Ayneband, 2008. Evaluation of bean (Phaseolus vulgaris) seeds inoculation with Rhizobium phaseoli and plant growth promoting rhizobacteria on yield and yield components. Pak. J. Biol. Sci., 11: 1935-1939.

Chapter 6

A Review of Organic Farming for Sustainable Agriculture in Northern India

S. K. Yadav, Subhash Babu, M. K. Yadav, Kalyan Singh, G. S. Yadav, and Suresh Pal

Institute of Agricultural Sciences, Banaras Hindu University, Varanasi, Uttar Pradesh 221005, India

ABSTRACT

In the post-independence period, the most important challenge in India has been to produce enough food for the growing population. Hence, high-yielding varieties are being used with infusion of irrigation water, fertilizers, or pesticides. This combination of high-yielding production technology has helped the country develop a food surplus as well as contributing to concerns of soil health, environmental pollution, pesticide toxicity, and sustainability of agricultural production. Scientists and policy planners are, therefore, reassessing agricultural practices which relied more on biological inputs rather than heavy usage of chemical fertilizers and pesticides. Organic farming can provide quality food without adversely affecting the soil's health and the environment; however, a concern is whether large-scale organic farming will produce enough food for India's large population. Certified organic products including all varieties of food products including basmati rice, pulses, honey, tea, spices, coffee, oilseeds, fruits, cereals, herbal medicines, and their value-added products are produced in India. Non edible organic products include cotton, garments, cosmetics, functional food products, body care products, and similar products. The production of these organic crops and products is reviewed with regard to sustainable agriculture in northern India.

INTRODUCTION

The organic movement in India has its origin in the work of Howard [1] who formulated and conceptualized most of the views which were later accepted by those people who became active in this movement. Organic farming is a production system which avoids, or largely excludes, the use of synthetic fertilizers, pesticides, growth regulators, and livestock feed additives. The objectives of environmental, social, and economic sustainability are the basics of organic farming [2]. The key characteristics include protecting the long-term fertility of soils by maintaining organic matter levels, fostering soil biological activity, careful mechanical intervention, nitrogen self-sufficiency through the use of legumes and biological nitrogen fixation, effective recycling of organic materials including crop residues and livestock wastes and weed, and diseases and pest control relying primarily on crop rotations, natural predators, diversity, organic manuring, and resistant varieties. A great emphasis is placed to maintain the soil fertility by returning all the wastes to it chiefly through compost to minimize the gap between NPK addition and removal from the soil [3]. Today, the burgeoning population pressure has forced many countries to use chemicals and fertilizers to increase the farm productivity for meeting their ever-increasing food requirements. The prolonged and over usage of chemicals has, however, resulted in human and soil health hazards along with environmental pollution. Farmers in the developed countries are, therefore, being encouraged to convert their existing farms into organic farm.

The key factors affecting consumer demand for organic food is the health consciousness and the willingness of the public to pay for the high-priced produce. In general, consumers of organic products are an affluent, educated, and health conscious group spurred by strong consumer demand, generous price premium, and concerns about the environment. Because of these hidden benefits, conventional growers are turning to organic farming. In Europe, government policies aim to stimulate the organic sector through subsidies, consumer education, and support in the form of research, education, and marketing. Agricultural practices of India date back to more than 4000 years, and organic farming is very much native to this country. As mentioned inArthashastra, farmers in the Vedic period possessed a fair knowledge of soil fertility, seed selection, plant protection, sowing seasons, and sustainability of crops in different lands [4]. The farmers of ancient India adhered to the natural laws and this helped in maintaining the soil fertility over a relatively longer period of time [5].

ORGANIC SOURCES OF PLANT NUTRIENTS

At present, most optimistic estimates show that about 25–30 percent of nutrient needs of Indian agriculture can be met by various organic sources. Supplementation of entire N through FYM sustains crop productivity at more than use of conventional N fertilizers. Since the estimates of NPK availability from organic sources are based on total nutrient content, efficiency of these sources to meet the nutrient requirement of crops is not as assured as mineral fertilizers, but the joint use of chemical fertilizers along with various organic sources is capable of sustaining higher crop productivity, improving soil quality, and productivity on long-term basis [3]. These organic sources besides supplying N, P, and K also make unavailable sources of elemental nitrogen, bound phosphates, micronutrients, and decomposed plant residues into an available form to facilitate the plants to absorb the nutrients. Application of organic sources encouraged the growth and activity of mycorrhizae and other beneficial organisms in the soil and is also helpful in alleviating the increasing incidence or deficiency of secondary and micronutrients and is capable of sustaining high crop productivity and soil health [6]. The farmers can in turn, get good remuneration from organically produced crops and if included in high value crop rotations, that is, aromatic rice (Oryza sativa L.), table pea (Pisum sativum L.), and onion (Allium cepa L.) [7] due to their heavy demands in domestic, national, and international markets.

Nutrient concentrations in FYM are usually small and vary greatly depending upon source, conditions, and duration of storage. The N, P, and K contents of fresh FYM range widely from 0.01 to 1.9 percent on dry weight basis due to variable nature of manure production and storage [8, 9]. Tandon [10] reported that on an average, well-rotted FYM contains 0.5 per cent N, 0.2 per cent P_2O_5, and 0.5 per cent K_2O. Gaur [11] stated that an application of 25 t ha^{-1} of well-rotted FYM can add 112 kg N, 56 kg P_2O_5, and 112 kg K_2O ha^{-1}. Several researchers all over the world have shown various benefits of the application of FYM on soil properties and productivity of crops [12]. Farmers generally use straw of the harvested crop as animal feed or bedding. In most cases, straw is used as bedding to trap urine to increase N cycling. Wet straw and manures from the animal sheds are collected every day and stored or composted on the farmer's premises. The composted manure is applied either immediately or stored until the next crop season depending upon farmer's socioeconomic conditions. In particular, soil, water, and nutrient management strategies, such as reduced tillage and use of raised beds, that avoid the deleterious effects of puddling on soil structure and fertility, improve water- and nutrient-use efficiencies, and increase crop productivity, may be appropriate [13].

EFFECT OF ORGANIC NUTRITION ON CROP PRODUCTIVITY

Addition of organic matter in the soil is a well-known practice to increase crop yields. Sharma and Mitra [14] reported that the application of organic materials increased grain and straw yield of rice. Ranganathan and Selvaseelan [15] found that application of spent mushroom and rice straw compost though comparable with FYM increased rice grain yields by 20 per cent over NPK fertilizer. Singh et al. [16] reported that the application of 7.5 t FYM ha^{-1} produced significantly more grain, and straw yields over unfertilized fields. All of the yield attributing characters of rice increased with increasing rates of FYM. Organic farming with dhaincha (Sesbania aculeata L.) made considerable improvement in grain yield of rice and Chickpea [17, 18]. Stockdale et al. [2] narrated the benefits of organic farming to developed nations (environmental protection, biodiversity enhancement, and reduced energy use and CO_2 emissions) and to developing countries (sustainable resources use, increased crop yield without over reliance on costly inputs, and environmental and biodiversity protection).

Many researchers reported that in an organically managed field activity of earth worm is higher than in inorganic agriculture [19]. In the biodegradation process earthworms and microbes work together and produce vermicompost, which is the worm fecal matter with worm casts. Vermicompost provided macroelements such as N, P, K, Ca, and Mg and microelements such as Fe, Mo, Zn, and Cu [20]. The vermicompost contained 0.74, 0.97, and 0.45 per cent nitrogen, phosphorus, and potassium, respectively [21].

In low-input agriculture, the crop productivity under organic farming is comparable to that under conventional farming. Tamaki et al. [22] reported that the growth of rice was better under continuous organic farming than with conventional farming. Agroeconomic study of practices of growing maize with compost and liquid manure top dressing in low-potential areas showed significantly better performance than those of current conventional farmer practices of a combined application of manure and mineral fertilizers. Maize grain yields were 11–17 per cent higher than those obtained with conventional practices [23].

Productivity of the crop during the initial year in an organically managed field is lower than in subsequent years as soil fertility levels increase over time as organic materials are added in the organic management system [24]. Similarly, Surekha [25] revealed that a gradual increase in grain yield with the use of organic fertilizers over a period of time was observed. Chan et al. [26] showed that the input of organic rice production in three different regions was 46, 25, and 22 per cent higher than conventional rice production, but rice yield was only 55, 94, and 82 per cent of conventional rice production, respectively. However, the cost of lower yield with higher inputs is compensated by the higher premium prices of organically crops in the markets [26].

Vegetables are highly responsive to organic sources of nutrients and profitable to farmers. Kalembasa [27] reported that vermicompost application of 15 kg per square meter gave the highest yield in tomato crop. Singh et al. [28] studied the response of chilli (Capsicum annuum L.) to vermicompost and observed that the application of vermicompost increased the microbial activities. Vermicompost has a positive effect on the performance of crops due to a higher number of branches and fruits [28]. Tomar et al. [29] recorded the highest yield (97 g $plant^{-1}$) through vermicompost in brinjal (Solanum melongena L.). Kalembasa and Deska [30] obtained significantly higher yield of sweet pepper (Capsicum annum L. var. grossum) with vermicompost. Reddy et al. [31] recorded maximum plant height at harvest, days to first flowering, and branches $plant^{-1}$ with the application of vermicompost (10 t ha^{-1}). Similarly, Tomar et al. [29] reported that the application of vermicompost significantly increased leaf area in carrot (Daucus carota L.) plants.

Manjarrez et al. [32] conducted an experiment on chili receiving 1.25, 2.0, 3.0, 4.0, 6.0, or 10.0 g of vermicompost kg^{-1} of soil under greenhouse conditions and reported that the foliar area and photosynthetic rate rose with increasing vermicompost application, and the highest photosynthetic rate (12 mol CO_2 m^{-2} s^{-1}) was observed with vermicompost at 10 g kg^{-1} soil. Atiyeh et al. [33] observed that when 20 per cent commercial horticultural medium was replaced by vermicompost there was significant increase in plant height and root and shoot biomass in tomato crop. Ribeiro et al. [34] observed that dry matter content increased with increasing the vermicompost dose upto 400 g kg^{-1} soil in sweet pepper cv. Nacional Ag. 506. Atiyeh et al. [33] conducted an experiment in which tomatoes were grown in a standard commercial greenhouse container medium (Metro-Mix 360, Manufacturer: Sun Gro Horticulture Canada Ltd., 770 Silver Street Agawam, MA, USA, 01001), considered as control, substituted with 10, 20, 30, 40, 50, 60, 70, 80, 90, and 100 per cent (by volume) pig (Sus scrofa L.) manure vermicompost. They obtained highest marketable yield (5.1 kg per plant) with substitution of Metro-Mix 360 with 20 per cent vermicompost. Substitution of Metro-Mix 360 with 10, 20, and 40 per cent vermicompost reduced the proportion of fruit that were nonmarketable and produced more large size (diameter 6.4 cm) than small size (diameter 5.8 cm) fruits. Shreeniwas et al. [35] conducted a field experiment on ridge gourd (Luffa acutangula L. Roxb.) and observed that the increasing levels of vermicompost (0, 5, 10, and

15 t ha^{-1}) increased the fruit weight and fruit volume. Rao and Sankar [36] observed that the effect of organic manure on leaf number, leaf area index, dry matter production, and other growth characters was significantly better than those of inorganic fertilizer in brinjal.

Samawat et al. [37] reported that vermicompost had a significant effect on root and fruit weight of tomatoes. In 100 percent vermicomposted treatment, fruit, shoot, and root weights were three, five, and nine times, respectively more than control. Where vermicompost was applied at at 5 t ha^{-1} or at 10 t ha^{-1}, increased shoot weight and leaf area of pepper plants (Capsicum annuum L.) compared to inorganic fertilizers [38]. Choudhary et al. [39] obtained the highest yield and available N of tomato cv. S-22 and cabbage (Brassica oleracea L.var. capitata) cv. Golden Acre with vermicompost at 200 g/plant + FYM at 250 g/plant, while maximum K and soil organic carbon was obtained with vermicompost at the rate of 100 g $plant^{-1}$ + FYM at 500 g $plant^{-1}$. Hashemimajd et al. [40] revealed that the treatment vermicompost produced from raw dairy manure (RDM) along with some other compost (sewage sludge + rice hull) assimilated higher shoot and root dry matter (DM) of tomatoes than the control (soil + sand).

Patil et al. [41] reported that total potato (Solanum tuberosum L.) tubers yield was significantly higher with the application of vermicompost at 4 t ha^{-1} and FYM at 25 t ha^{-1}. Sawicka et al. [42] reported that the cultivation system had the strongest effect on the share of commercial potato tubers and tubers of a diameter of 4–6 cm in the total yield. Haase et al. [43] suggested that tubers from organic potato cropping may be expected to have sufficiently high tuber dry matter concentrations (19%) for processing into French fries without impairing the texture of the fries when concentration exceeds 23%. Dry matter concentration of tubers for crisps (cv. Marlen) fell short of the required minimum of 22% when a combined N and K fertilizer was applied. Mourao et al. [44] found that organically grown potato cv. Virgo yielded 66% of the conventional crop, whereas Raja yielded 46.6%. The nitrogen uptake of organic crop (tubers and foliage) was 37.0 kg/ha for Raja and 50.5 kg/ha for Virgo compared to that of 21.1% and 27.8% of nitrogen uptake, respectively, with mineral fertilizer.

Addition of organic amendments and casting of earthworms to soil also proved effective in controlling diseases in pea (Pisum sativum L.), mustard (Brassica juncea L. Coss.), and chickpea (Cicer arietinum

L.) during winter season. Nitrogen, phosphorus, potassium, calcium, and magnesium accumulation also increased with increasing doses of vermicompost as well as with fertilizers [45]. Singh [46] observed that the application of vermicompost at 13–20 q ha^{-1} increased yield of pea (23.62 q ha^{-1}) and groundnut (Arachis hypogaea L.) (12.16 q ha^{-1}). The principal findings of Jat and Ahlawat [47] revealed that the application of 3 t vermicompost ha^{-1} to chickpea improved dry matter accumulation, grain yield, and grain protein content in chickpea, soil nitrogen and phosphorus and bacterial count, dry fodder yield of succeeding maize (Zea mays L.), and total nitrogen and phosphorus uptake by the cropping system over no vermicompost. Baswana and Rana [48] reported that the highest pod yield (93.96 q/ha) of pea was recorded when farm yard manure (1 t ha^{-1}) + poultry manure (1 t ha^{-1}) along with mulch treatment was applied followed by farm yard manure (2 t ha^{-1}) + biofertilizers with mulch treatment. Similar trend was also observed for biological yield and harvest index.

Dayal and Agarwal [49] observed that the seed yield of sunflower (Helianthus annus L.) was increased with the higher rate of vermicompost (10 t ha^{-1}); the best combination was 5 t ha^{-1} vermicompost. Somasundaram et al. [50] reported that the study revealed that increased soluble protein content and nitrogenase activity of maize, sunflower, and green gram (Vigna radiata L.) was estimated with biogas slurry. Increased nitrogen accumulation at all growth stages on maize, sunflower, and green gram was observed under biogas slurry with panchagavya. Higher yield of maize and sunflower was recorded under biogas slurry with panchagavya (a preparation of 5 cow products (dung, urine, milk, ghee and curds)). Silwana et al. [51] reported the importance of organic manure and its long time usefulness in increasing productivity of maize-bean (Phaseolus vulgaris L.) intercrop for small-scale farmers in Eastern Cape of South Africa.

Sangakkara et al. [52] found that the organic matter incorporation increased soil water retention in soil and hence enhanced root growth, culminating in high yields of maize. The impact was greater in maize than in cowpea, especially with gliricidia leaves. Seo and Lee [53] reported that soil organic nitrogen increased considerably by hairy vetch. Dry matter yields of maize increased more in hairy vetch than ammonium nitrate with N rates over 160 kg ha^{-1}. Adiku et al. [54] revealed that the fertilized maize-grass and maize-pigeon pea (Cajanus cajan L. Millspaugh) rotations were identified as those that sustained relatively high maize yields, returned large residue amounts to the soil, and minimized soil carbon loss.

Oliveira et al. [55] reported that the highest average head weight (700 g) and yield (38 t ha^{-1}) in cabbage cv. Matsukaze was produced with the application of earthworm compost at 27 and 29 t ha^{-1}, respectively. Datta et al. [56] confined that the inoculation of seed with Rhizobium leguminosarum bv. phaseoli and incorporation of FYM one week before sowing of rajmash (Phaseolus vulgaris L.) increased yield. Similarly, inoculation of seed enhanced N fixation and incorporation of FYM left a net positive balance of 42 and 84 kg N, respectively, with regards to control (no seed inoculation and no FYM incorporation in soil). A higher accumulation rate of available N at all the growth stage of rajmash was observed with incorporation of FYM and inoculation of seed over control (no seed inoculation and no FYM incorporation in soil).

In all four of the years studied, the organic and conventional farming systems did not show significant differences in marketable yields for any vegetable crops, namely, tomato, bean, cabbage, and zucchini (Cucurbita pepo L.). The yields in organic farming were 10 per cent and 3 per cent, respectively, higher than conventional farming [57].

Sarangthem and Salam [58] reported that the application of decomposed urban waste with total nitrogen 0.58–1.9 per cent, available phosphorus 0.45–0.67 per cent, and available potash 1.4–1.8 per cent increased the yield of bean to 228 gm/pot from 53 gm/pot. The response on growth and yield of bean (228 g/pot) was recorded higher in the decomposed manure enrich with vermiculture. Renuka and Sankar [59] reported in tomato that the yield increased two and half times with the application of organic manures in comparison with inorganic fertilizer (18.44 tonnes). Likewise, Samawat et al. [37] reported that vermicompost had a significant effect on the number of fruits in tomato. In 100 per cent vermicomposted treatment, fruit numbers were four times more than the control treatment. Arancon et al. [38] reported that when vermicompost applied at 5 t ha^{-1} or 10 t ha^{-1}, the marketable tomato yield in all vermicompost treated plots were considerably greater than yield from the inorganic fertilizer plots. The total and marketable fruit yield of pepper also increased with vermicompost compared with inorganic fertilizers. Thanunathan et al. [60] reported that soil + mine spoil + coir pith vermicompost (1 : 1 : 1) significantly increased plant height, number of leaf, and root length in onion (Allium cepa L.). Lopes et al. [61] reported that the application of vermicompost at 10 t ha^{-1} significantly increased nodulation and dry matter yield of cowpea (Vigna sinensis L.) over its lower levels, namely, 0 and 5 t ha^{-1}.

EFFECT OF ORGANIC NUTRITION ON QUALITY PARAMETERS OF CROPS

Yadav and Vijayakumari [62] carried out an experiment to assess the effect of vermicomposted vegetable waste on the biochemical characters of chilli and found that the protein was higher at 60 (113 mg g^{-1}) and 90 DAS (79 mg g^{-1}). The carbohydrate content was higher in vermicomposted treatment at 60 DAS (15.34 mg g^{-1}). Chlorophyll (2.61 mg g^{-1}) and total chlorophyll (3.62 mg g^{-1}) contents were observed at 60 DAS, while chlorophyll a (1.01 mg g^{-1}) was higher at 90 DAS as compared to inorganic fertilizers. In another experiment, Haase et al. [43] suggested that tubers from organic potato cropping may be expected to have sufficiently high tuber dry matter concentrations (19 per cent) for processing into French fries without impairing the texture of the fries when concentrations exceed 23 per cent. Similarly, application of FYM at 10 t ha^{-1} alone increased the economic yield and quality parameters like hulling percentage, milling percentage, and protein and amylose content of rice cv. Saket-4 [63].

Mourao et al. [44] found that organically grown potatocv. Virgo yielded 66 per cent of the conventional crop, whereas Raja yielded 47 per cent. The nitrogen uptake of organic crop (tubers and foliage) was 37.0 kg/ha for Raja and 50.5 kg/ha for Virgo, respectively, 21 and 28 per cent of nitrogen uptake by same cultivars grown with mineral fertilizer. Although foliage nitrogen content was increased for the conventional crops, difference between N content of organic and conventional tubers were not significant, as well as for K, Ca, and Mg. Maheswari et al. [64] studied the effect of foliar organic fertilizers on the quality and economics of chilli and observed the highest ascorbic acid content (175.23 mg/100 g) with vermiwash : water at 1 : 5 ratio.

EFFECT OF ORGANIC NUTRITION ON SOIL FERTILITY

Minhas and Sood [65] also reported that the organic matter after decomposition release macro- and micronutrients to the soil solution, which becomes available to the plants, resulting in higher uptake. Organic farming was capable of sustaining higher crop productivity and improving soil quality and productivity by manipulating the soil properties on long term basis. It was reported that organic and low-input farming practices after 4 years led to an increase in the organic carbon, soluble phosphorus, exchangeable potassium, and pH and also the reserve pool of stored nutrients and maintained relativity stable EC level [66, 67].

Normal composting takes a long time leading to considerable loss of organic materials as CO_2 or does not contribute to the organic pool [68]. Bulluck et al. [69] reported that the use of compost raised soil pH from 6.0 without compost to 6.5 with compost and reduced the broadleaf weed population by 29 percent and grassy weed population by 78 per cent. Degradation of soil organic matter reduced nutrient supplying capacity, especially, on soils with high initial soil organic matter content in rice-wheat cropping system [70]. Organic farming improved organic matter content and labile status of nutrients [71] and also soil physicochemical properties. Addition of carbonaceous materials such as straw, wood, bark, sawdust, or corn cobs helped the composting characteristics of a manure. These materials reduced water content and raised the C:N ratio. However, under Indian conditions, joint composting of the manure slurries with plant residues was more viable and profitable than its separate composting. Use of FYM and green manure maintained high levels of Zn, Fe, Cu, and Mn in rice-wheat rotation [72].

Laxminarayana and Patiram [73] concluded that the decline in soil reaction might be due to organic compounds added to the soil in the form of green as well as root biomass which produced more humus and organic acids on decomposition. Urkurkar et al. [74] reported that supply of 100 per cent nitrogen, that is, 120 kg/ha for rice and 150 kg/ha for potato in a rice-potato cropping system 1/3 each from cow dung manure, neem cake, and composed crop residue appreciably increased the organic carbon (6.3 g kg^{-1}) over initial value (5.8 g kg^{-1})

as compared to supply from inorganic fertilizers alone. However, availability of phosphorus and potassium did not show any perceptible change after completion of five cropping cycles under organic as well as integrated nutrient approaches.

EFFECT OF ORGANIC NUTRITION ON SOIL BIOLOGICAL PROPERTIES

Compost contains bacterial, actinomycetes, and fungi; hence, a fresh supply of humic material not only added microorganisms but also stimulated them [75, 76]. Besides, compost played an important role in control of plant nematodes and in mitigating the effect of pesticides through sorption. Sorption is the most important interaction between soil/organic matter and pesticides and limits degradation as well as transport in soil. Pesticides bound to soil organic matter or clay particles are less mobile, bioavailable but also less accessible to microbial degradation and thus more persistent [77–79]. Composting material added plenty of carbon and thus increased heterotrophic bacteria and fungi in soil and further increased the activity of soil enzymes responsible for the conversion of unavailable to available form of nutrients. The application of FYM with rhizobium and coinoculation of PSB with rhizobium augmented soybean (Glycine max L. Merr.) production [80].

Agricultural practices have had an impact on soil biophysiochemical properties. Densities of bacteria, protozoa, nematodes, and arthropods in soils under organic farming were higher than under conventional farming [81]. Bulluck et al. [82] reported that organic fertility amendments enhanced beneficial soil microorganisms, reduced pathogen population, total carbon, and cation exchange capacity, and lowered down bulk densities, thus improved soil quality.

The National Academy of Agricultural Sciences (NAAS) recommended a holistic approach involving integrated nutrient management (INM), integrated pest management (IPM) for enhanced input use efficiency, and adoption of region specific promising cropping systems as an alternative organic farming strategy for India and to begin with the practice of organic farming should value crops like spices, medicinal plants, fruits, and vegetables [83].

Singh and Bohra [84] reported that rice-pea-black gram (Vigna mungo L.) cropping system recorded higher population of bacteria, actinomycetes, and fungi than rice-wheat cropping system. Field experiment conducted with P solubilizers like Aspergillus awamori, Pseudomonas striata, and Bacillus polymyxa significantly increased the yield of various crops like wheat, rice, cowpea (Vigna sinensis L.), and so forth in presence of rock phosphate and saved 30 Kg P_2O_5 ha^{-1} with the use of phosphate solubilizing microorganisms. Vegetable crops, in general, responded better to Azotobacter inoculation than other field crops. Nevertheless, yield increase in case of wheat, maize, jowar (Sorghum bicolor L. Moench), cotton (Gossypium spp.), and mustard crop usingAzotobacter chrooccocum culture was 0–31 per cent higher than control [85].

In low-input agriculture, the crop productivity under organic farming is comparable to conventional farming. Integrated use of rice straw compost + Azotobacter and PSB was found better than rice straw alone [86]. Azotobacter produced growth promoting substances which improved seed germination and growth with extended root system. It also produced polysaccharides which improved soil aggregation [87]. Seed inoculation of chickpea with rhizobium + PSB (phosphate solubilising bacteria) increased dry matter accumulation, grain yield, and grain protein content in chickpea, dry fodder yield of succeeding maize, and total nitrogen and phosphorus uptake by the cropping system over no inoculation and inoculation with rhizobium alone.

CONCLUSIONS

Organic farming can provide quality food without adversely affecting the soil's health and the environment. There is need to identify suitable crops/products on regional basis for organic production that has international market demands. The whole region as such cannot afford to go for organic at a time because of its commitments to insure food and nutritional security. This will provide ample opportunity for employment and bring prosperity and peace in the region.

REFERENCES

1. A. Howard, An Agricultural Testaments, Oxford University Press, 1940.
2. E. A. Stockdale, N. H. Lampkin, M. Hovi et al., "Agronomic and environmental implications of organic farming systems," Advances in Agronomy, vol. 70, pp. 261–327, 2001.
3. P. K. Chhonkar, "Organic farming myth and reality," in Proceedings of the FAI Seminar on Fertilizer and Agriculture Meeting the Challenges, New Delhi, India, December 2002.
4. P. K. Sofia, R. Prasad, and V. K. Vijay, "Organic farming-tradition reinvented," Indian Journal of Traditional Knowledge, vol. 5, no. 1, pp. 139–142, 2006.
5. S. Chandra and S. K. Chauhan, "Prospects of organic farming in India," Indian Farming, vol. 52, no. 2, pp. 11–14, 2004.
6. K. K. M. Nambiar, P. N. Soni, M. R. Vats, D. K. Sehgal, and D. K. Mehta, "AICRP on long term fertilizer experiments," Annual Reports 1987-88 and 1988-89, IARI, New Delhi, India, 1992.
7. S. Kalyan, "Development of sustainable farming system model for the Irrigated agro-ecosystem of Eastern UP, ICAR, Ad-hoc project," Final Annual Report, Department of Agronomy, Institute of Agricultural Science, Banaras Hindu University, Varanasi, India, 2005.
8. A. Inoko, "Compost as source of plant nutrients," in Organic Matter and Rice, S. Banta and C. V. Mendoza, Eds., pp. 137–146, IRRI, Los Banos, Philippines, 1984.
9. Z. I. Zhu, C. Q. Liu, and B. F. Jiang, "Mineralization of organic nitrogen, phosphorus and sulphur in some paddy soils in China," in Organic Matter and Rice, pp. 259–272, IRRI, Los Banos, Philippines, 1984.
10. H. L. S. Tandon, "Fertilizers and their integration and organics and bio-fertilizers," in Fertilizers, Organic Manures, Recyclable Wastes and Bio-Fertilizers, H. L. S. Tandon, Ed., pp. 32–36, FDCO, New Delhi, India, 1992.
11. A. C. Gaur, "Bulky organic manures and crop residues," in Fertilizer Organic Manures Recyclable Wastes and Bio-Fertilizers, H. L. S. Tandon, Ed., Fertiliser Development and Consultation Organisation, New Delhi, India, 1992.

12. R. T. Prabhakar, M. Umadevi, and R. P. Chandrasekhar, "Effect of fly ash and farm yard manure on soil properties and yield of rice grown on an inceptisol," Agricultural Science Digest, vol. 30, no. 4, pp. 281–285, 2010.
13. J. Timsina and D. J. Connor, "Productivity and management of rice-wheat cropping systems: issues and challenges," Field Crops Research, vol. 69, no. 2, pp. 93–132, 2001.
14. A. R. Sharma and B. N. Mitra, "Complementary effect of organic material in rice-wheat crop sequence,"The Indian Journal of Agricultural Sciences, vol. 60, no. 3, pp. 163–168, 1990.
15. D. S. Ranganathan and D. A. Selvaseelan, "Mushroom spent rice straw compost and composted coir pith as organic manures for rice," Journal of the Indian Society of Soil Science, vol. 45, no. 3, pp. 510–514, 1997.
16. G. Singh, O. P. Singh, R. A. Yadava, P. P. Singh, and R. Nayak, "Response of rice (Oryza sativa) varieties to nitrogen levels in flash flood conditions," Indian Journal of Agronomy, vol. 43, no. 3, pp. 506–510, 1998.
17. K. N. Singh, B. Prasad, and S. K. Sinha, "Effect of integrated nutrient management on a Typic Haplaquant on yield and nutrient availability in a rice-wheat cropping system," Australian Journal of Agricultural Research, vol. 52, no. 8, pp. 855–858, 2001.
18. K. N. Singh, I. P. Sharma, and V. C. Srivastava, "Effect on FYM fertilizer and plant density on productivity of rice-what sequence," Journal of Research, Birsa Agricultural University, vol. 13, no. 2, pp. 159–162, 2001.
19. C. A. Edwards and J. R. Lofty, "The invertebrate fauna of the Park Grassplots. I: soil fauna," Rothamsted Report, part 2, pp. 133–154, 1974.
20. K. Amir and I. Fouzia, "Chemical nutrient analysis of different composts (Vermicompost and Pitcompost) and their effect on the growth of a vegetative crop Pisum sativum," Asian Journal of Plant Science and Research, vol. 1, no. 1, pp. 116–130, 2011.
21. M. Pal, Basics of Agriculture, Jain Brothers, New Delhi, India, 2002.
22. M. Tamaki, T. Itani, and H. Nakano, "Effects of organic ad inorganic fertilizers on the growth of rice plants of rice plants

under different light intensities," Japanese Journal of Crop Science, vol. 71, no. 4, pp. 439–445, 2002.

23. D. D. Onduru, J. M. Diop, E. Van der Werf, and A. De Jager, "Participatory on-farm comparative assessment of organic and conventional farmers' practices in Kenya," Biological Agriculture and Horticulture, vol. 19, no. 4, pp. 295–314, 2002.
24. S. K. Yadav, S. Yogeshwar, M. K. Yadav, B. Subhash, and S. Kalyan, "Effect of organic nitrogen sources on yield, nutrient uptake and soil health under rice (Oryza sativa) based cropping sequence," Indian Journal of Agricultural Sciences, vol. 83, no. 2, pp. 170–175, 2013.
25. K. Surekha, "Nitrogen-release pattern from organic sources of different C:N ratios and lignin content, and their contribution to irrigated rice (Oryza sativa)," Indian Journal of Agronomy, vol. 52, no. 3, pp. 220–224, 2007.
26. K. Y. Chan, C. Dorahy, T. Wells et al., "Use of garden organic compost in vegetable productionunder contrasting soil P status," Australian Journal of Agricultural Research, vol. 59, no. 4, pp. 374–382, 2008.
27. D. Kalembasa, "The effects of vermicompost on the yield and chemical composition of tomato," Zeszyty Problemowe Postępów Nauk Rolniczych, vol. 437, pp. 249–252, 1996.
28. J. Singh, S. S. Bhat, M. R. Sudharshan, Jasvir-Singh, and S. Sreekrishna-Bhat, "Performance of Scotch Bonnet chilli in Karnataka and its response to vermicompost," Indian Cocoa, Arecanut & Spices Journal, vol. 21, pp. 9–10, 1997.
29. V. K. Tomar, R. K. Bhatnagar, and R. K. Palta, "Effect of vermicompost on production of brinjal and carrot," Bhartiya Krishi Anusandhan Patrika, vol. 13, no. 3-4, pp. 153–156, 1998.
30. S. Kalembasa and J. Deska, "The possibility of utilizing vermicompost in the cultivation of radish and paprika," Annals of Agricultural University of Poznan, vol. 27, pp. 131–136, 1998.
31. R. Reddy, M. A. N. Reddy, Y. T. N. Reddy, N. S. Reddy, and M. Anjanappa, "Effect of organic and inorganic sources of NPK on growth and yield of pea," Legume Research, vol. 21, no. 1, pp. 57–60, 1998.

32. Manjarrez, M. J. Martinez, R. Ferrera Cerrato, and M. C. Gonzalez Chavez, "Effect of vermicompost and mycorrhizal fungi on growth and photosynthetic rate of chilli," Terra, vol. 17, no. 1, pp. 9–15, 1999.
33. R. M. Atiyeh, N. Arancon, C. A. Edwards, and J. D. Metzger, "Influence of earthworm-processed pig manure on the growth and yield of greenhouse tomatoes," Bioresource Technology, vol. 75, no. 3, pp. 175–180, 2000.
34. L. G. Ribeiro, J. C. Lopes, F. S. Martins, and S. S. Ramalho, "Effect of organic fertilizer application on sweet pepper yield," Horticultura Brasileira, vol. 18, pp. 134–137, 2000.
35. C. H. Shreeniwas, S. Muralidhar, and M. S. Rao, "Yield and quality of ridge gourd fruits as influenced by different levels of inorganic fertilizers and vermicompost," Annals of Agricultural Research, vol. 21, pp. 262–266, 2000.
36. T. S. S. Rao and C. R. Sankar, "Effect of organic manures on growth and yield of brinjal," South Indian Horticulture, vol. 49, pp. 288–291, 2001.
37. S. Samawat, A. Lakzian, and A. Zamirpour, "The effect of vermicompost on growth characteristics of tomato," Agricultural Science and Technology, vol. 15, no. 2, pp. 83–89, 2001.
38. N. Q. Arancon, C. A. Edwards, P. Bierman, J. D. Metzger, S. Lee, and C. Welch, "Effects of vermicomposts on growth and marketable fruits of field-grown tomatoes, peppers and strawberries, the 7th international symposium on earthworm ecology, Cardiff, Wales, 2002," Pedobiologia, vol. 47, no. 5-6, pp. 731–735, 2003.
39. R. S. Choudhary, A. Das, and U. S. Patnaik, "Organic farming for vegetable production using vermicompost and FYM in Kokriguda watershed of Orissa," Indian Journal of Soil Conservation, vol. 31, no. 2, pp. 203–206, 2003.
40. K. Hashemimajd, M. Kalbasi, A. Golchin, and H. Shariatmadari, "Comparison of vermicompost and composts as potting media for growth of tomatoes," Journal of Plant Nutrition, vol. 27, no. 6, pp. 1107–1123, 2004.
41. M. P. Patil, N. C. Hulamani, S. I. Athani, and M. G. Patil, "Response of tomato (Salanum tuberosum) cv. Kufri Chandramukhi to integrated nutrient management," Advances in Agricultural Research in India, vol. 8, pp. 135–139, 1997.

42. B. Sawicka, P. Barbas, and J. Kus, "Variability of potato yield and its structure in organic and integrated crop production systems," Electronic Journal of Polish Agricultural Universities, vol. 10, no. 1, pp. 425–427, 2007.
43. T. Haase, C. Schüler, N. U. Haase, and J. Heß, "Suitability of organic potatoes for industrial processing: effect of agronomical measures on selected quality parameters at harvest and after storage," Potato Research, vol. 50, no. 2, pp. 115–141, 2007.
44. I. Mourao, L. M. Brito, and J. Coutinho, "Cultivating the-future based-on science, volume-1: organic crop-production," in Proceedings of the 2nd Scientific Conference of the International Society of Organic Agriculture Research ISOFAR, held at the 16th IFOAM Organic World Conference in Cooperation with the International Federation of Organic Agriculture-Movements IFOAM and the Consorzio ModenaBio, pp. 596–599, International Society of Organic Agricultural Research (ISOFAR), Bonn, Germany, June 2008.
45. S. C. Tripathi, D. S. Chauhan, R. K. Sharma, and O. P. Dhillon, "Productivity and economics of different wheat (Triticum aestivum) based cropping sequences," Indian Journal of Agronomy, vol. 44, no. 2, pp. 237–241, 1999.
46. R. A. Singh, "Feasibility of vermi-farming in peanut (Arachis hypogaea)-vegetable pea (Pisum sativum) cropping system," Indian Journal of Agronomy, vol. 45, no. 2, pp. 257–262, 2000.
47. R. S. Jat and I. P. S. Ahlawat, "Direct and residual effect of vermicompost, biofertilizers and phosphorus on soil nutrient dynamics and productivity of chickpea-fodder maize sequence," Journal of Sustainable Agriculture, vol. 28, no. 1, pp. 41–54, 2006.
48. K. S. Baswana and M. K. Rana, "Effect of organic sources and biofertilizers on growth and yield of garden pea (Pisum sativum L)," Haryana Journal of Horticultural Sciences, vol. 36, no. 3-4, pp. 326–330, 2007.
49. D. Dayal and S. K. Agarwal, "Response of sunflower genotypes (Helianthus annuus) to nutrient management," Indian Journal of Agricultural Sciences, vol. 69, no. 1, pp. 10–13, 1999.
50. E. Somasundaram, M. A. Mohamed, K. Thirukkumaran, Chandrasekaran, K. Vaiyapuri, and K. Sathyamoorthi,

"Biochemical changes, nitrogen flux and yield of crops due to organic sources of nutrients under maize based cropping system," Journal of Applied Sciences Research, vol. 3, pp. 1724–1729, 2007.

51. T. T. Silwana, E. O. Lucas, and A. B. Olaniyan, "The effects of inorganic and organic fertilizers on the growth and development of component crops in maize/bean intercrop in Eastern Cape of South Africa,"Journal of Food, Agriculture and Environment, vol. 5, no. 1, pp. 267–272, 2007.

52. R. Sangakkara, P. Bandaranayake, U. Dissanayake, and J. N. Gajanayake, "Organic matter addition in organic farming—impact on root development and yields in maize and cowpea over dry seasons," inProceedings of the 2nd Scientific Conference of the International Society of Organic Agriculture Research (ISOFAR ‹08), held at the16th IFOAM Organic World Conference in Cooperation with the International Federation of Organic-Agriculture Movements IFOAM and the Consorzio ModenaBio, Modena, Italy, June 2008.

53. J. Seo and H. J. Lee, "Mineral nitrogen effects of hairy vetch (Vicia villosa Roth) on maize (Zea mays L.) by green manure amounts," Journal of Agronomy, vol. 7, no. 3, pp. 272–276, 2008.

54. S. G. K. Adiku, J. W. Jones, F. K. Kumaga, and A. Tonyigah, "Effects of crop rotation and fallow residue management on maize growth, yield and soil carbon in a savannah-forest transition zone of Ghana,"Journal of Agricultural Science, vol. 147, no. 3, pp. 313–322, 2009.

55. A. P. Oliveira, D. S. Ferreira, C. C. Casta, A. F. Silva, and E. V. Alves, "Utilization of cattle manure and earthworm compost in hybrid cabbage production," Horticulture Brasileira, vol. 13, pp. 70–73, 2001.

56. N. Datta, M. C. Rana, and R. P. Sharma, "Effect of seed inoculation and farm yard manuring on nitrogen balance and yield in Rajmash (Phaseolus vulgaris)," Indian Journal of Plant Physiology, vol. 11, no. 1, pp. 108–112, 2006.

57. V. Tabaglio, C. Gavzzi, and G. Nervo, "Cultivating the future based on science. Volume 1: organic crop production," in Proceedings of the 2nd Scientific Conference of the International Society of Organic Agriculture Research, held at the 16th IFOAM

Organic World Conference in Cooperation with the International Federation of Organic Agriculture Movements (IFOAM) and-the-Consorzio ModenaBio, pp. 328–331, International Society of Organic Agriculture Research, Modena, Italy, June 2008.

58. I. Sarangthem and J. S. Salam, "Recycling of the urban waste as manure on French bean (Phaseolus vulgaris) as test crop," Indian Journal of Agricultural Research, vol. 42, no. 2, pp. 124–127, 2008.

59. B. Renuka and C. R. Sankar, "Effect of organic manures on growth and yield of tomato," South Indian Hort, vol. 49, pp. 216–219, 2001.

60. K. Thanunathan, S. Natarajan, R. Sentikumar, and K. Asulmurogan, "Effect of different sources of organic amendments on growth and yield of onion in mine spoil," Madras Agricultural Journal, vol. 84, no. 7, pp. 382–384, 1997.

61. A. J. Lopes, N. P. Stamford, M. V. B. Figueired, N. A. Burity, and E. B. Ferraz, "Effect of vermicompost, mineral nitrogen and mineralizing agent on N fixation and yield in cowpea," Revista Brasileira de Ciência to Solo, vol. 20, pp. 55–62, 1996.

62. R. H. Yadav and B. Vijayakumari, "Impact of vermicompost on biochemical characters of Chilli (Capsicum annum)," Journal of Ecotoxicology and Environmental Monitoring, vol. 14, no. 1, pp. 51–56, 2004.

63. K. G. Dixit and B. R. Gupta, "Effect of farmyard manure, chemical and bio-fertilizers on yield and quality of rice (Oryza sativa L.) and soil properties," Journal of the Indian Society of Soil Science, vol. 48, no. 4, pp. 773–780, 2000.

64. T. U. Maheswari, K. Haripriya, P. Poonkodi, and S. Kamala kannan, "Effect of foliar application of organic nutrients on some quality indices and economics of chilli (Capsicum annum L," Advances in Plant Sciences, vol. 17, no. 1, pp. 259–262, 2004.

65. R. S. Minhas and A. Sood, "Effet of inorganic and organic on yield and nutrients uptake by three crops in rotation in aid alfisol," Journal of the Indian Society of Soil Science, vol. 42, pp. 27–260, 1994.

66. M. S. Clark, W. R. Horwath, C. Shennan, and K. M. Scow, "Changes in soil chemical properties resulting from organic and

low-input farming practices," Agronomy Journal, vol. 90, no. 5, pp. 662–671, 1998.

67. A. C. Gaur, S. Nilkantan, and K. S. Dargan, Organic Manures, ICAR, New Delhi, India, 2002.
68. N. S. Subba Rao, "Organic matter decomposition," in Soil Microbiology, pp. 255–270, Oxford & IBH Publishing, New Delhi, India, 1999.
69. L. R. Bulluck, M. Brosius, G. K. Evanylo, and J. B. Ristaino, "Organic and synthetic fertility amendments influence soil microbial, physical and chemical properties on organic and conventional farms," Applied Soil Ecology, vol. 19, no. 2, pp. 147–160, 2002.
70. R. L. Yadav, B. S. Dwivedi, and P. S. Pandey, "Rice-wheat cropping system: assessment of sustainability under green manuring and chemical fertilizer inputs," Field Crops Research, vol. 65, no. 1, pp. 15–30, 2000.
71. S. Subbiah and K. Kumaraswamy, "Effect of manure-fertilizers on the yield and quality of rice and on soil fertility," Fertilizer News, vol. 45, no. 10, pp. 61–68, 2000.
72. S. P. Singh, R. M. Kumar, and S. V. Subbaiah, "Effect of nitrogen and potassium application on grain yield of hybrid rice (Oryza sativa)," in Proceedings of the 2nd International Agronomy Congress, vol. 1, New Delhi, India, 2002.
73. K. Laxminarayana and Patiram, "Effect of integrated use of inorganic, biological and organic manures on rice productivity and soil fertility in ultisols of Mizoram," Journal of Indian Society of Soil Science, vol. 54, no. 2, pp. 213–220, 2006.
74. J. S. Urkurkar, S. Chitale, and A. Tiwari, "Effect of organic v/s chemical nutrient packages on productivity, economics and physical status of soil in rice (Oryza sativa)-Potato (Solanum tuberosum) cropping system in Chhattisgarh," Indian Journal of Agronomy, vol. 55, no. 1, pp. 6–10, 2010.
75. A. Balasubramanian, R. Siddaramappa, and G. Rangaswami, "Effect of organic manuring on the activities of the enzymes hydrolysing sucrose and urea and on soil aggregation," Plant and Soil, vol. 37, no. 2, pp. 319–328, 1972.

76. A. C. Gaur, K. V. Sadasivam, O. P. Vimal, R. S. Mathur, and S. K. Kavimandan, "Studies on the humification of organic matter in a red Rakar soil," Zentralblatt fur Bakteriologie, vol. 128, no. 1, pp. 149–161, 1973.
77. A. C. Gaur and S. K. Prasad, "Effect of organic matter and inorganic fertilizers on plant parasitic nematodes," Indian Journal of Entomology, vol. 32, pp. 186–188, 1970.
78. S. K. Prasad, S. D. Mishra, and A. C. Gaur, "Effect of soil amendments on nematodes associated with wheat followed by mung and maize," Indian Journal of Entomology, vol. 34, pp. 307–311, 1972.
79. A. C. Gaur, "All Indian coordinated project on microbiological decomposition and recycling of farm and city wastes," Project Report, Indian Council of Agricultural Research, Poona, India, 1975.
80. K. N. Sharma and K. N. Namdeo, "Effect of bio-fertilizers and phosphorus on growth and yield of Soybean (Glycine max L. Merill)," Crop Research, vol. 17, no. 2, pp. 160–163, 1999.
81. S. Wu, E. R. Ingham, and D. Hu, "Soil microfloral and faunal populations in an organic ecosystem in Oregon, USA," in Proceedings of the 17th World Congress of Soil Science, vol. 5, p. 1756, Queen Sirkit National Convention Centre, Bangkok, Thailand, August 2002.
82. L. R. Bulluck III, M. Brosius, G. K. Evanylo, and J. B. Ristaino, "Organic and synthetic fertility amendments influence soil microbial, physical and chemical properties on organic and conventional farms," Applied Soil Ecology, vol. 19, no. 2, pp. 147–160, 2002.
83. P. Bhattacharya and G. Chakraborty, "Current status of organic farming in India and other countries,"Indian Journal of Fertilizers, vol. 1, no. 9, pp. 111–123, 2005.
84. K. Singh and J. S. Bohra, "Net-working project on diversification of rice wheat system through pulses and oilseeds," Project Report, UPCAR, 2009.
85. S. T. Shende and R. Apte, "Azospirillum inoculation—a highly remunerative input for agriculture," inProceedings of the National Symposium on Biological Nitrogen Fixation, pp. 532–541, New Delhi, India, February 1982.

86. S. Kalyan, "Integrated use of industrial and farm wastes with mineral and bio-activators for sustained rice productivity in rice based eco-system of Eastern Uttar Pradesh," First Annual Report, NATP, ICAR, New Delhi, India, 2003.
87. A. C. Gaur, Handbook of Organic Farming and Biofertilizers, Ambica Book Publication, Jaipur, India, 2006.

Chapter 7

Phosphate Solubilizing Microbes: Sustainable Approach for Managing Phosphorus Deficiency in Agricultural Soils

Seema B Sharma[1], Riyaz Z Sayyed[2], Mrugesh H Trivedi[1], and Thivakaran A Gobi[3]

[1]Department of Earth and Environmental Science, KSKV Kachchh University, Mundra Road, Bhuj 370 001, Gujarat, India

[2]Department of Microbiology, PSGVP Mandal's Arts, Science and Commerce College, Shahada 425409, Maharashtra, India

[3]Gujarat Institute of Desert Ecology, Bhuj, Gujarat, India

ABSTRACT

Phosphorus is the second important key element after nitrogen as a mineral nutrient in terms of quantitative plant requirement. Although abundant in soils, in both organic and inorganic forms, its availability is restricted as it occurs mostly in insoluble forms. The P content in average soil is about 0.05% (w/w) but only 0.1% of the total P is available to plant because of poor solubility and its fixation in soil (Illmer and Schinner, Soil Biol Biochem 27:257-263, 1995). An adequate supply of phosphorus during early phases of plant development is important for laying down the primordia of plant reproductive parts. It plays significant role in increasing root ramification and strength thereby imparting vitality and disease resistance capacity to plant. It also helps in seed formation and in early maturation of crops like cereals and legumes. Poor availability or deficiency of phosphorus (P) markedly reduces plant size and growth. Phosphorus accounts about 0.2 - 0.8% of the plant dry weight.

To satisfy crop nutritional requirements, P is usually added to soil as chemical P fertilizer, however synthesis of chemical P fertilizer is highly energy intensive processes, and has long term impacts on the environment in terms of eutrophication, soil fertilility depletion, carbon footprint. Moreover, plants can use only a small amount of this P since 75–90% of added P is precipitated by metal–cation complexes, and rapidly becomes fixed in soils. Such environmental concerns have led to the search for sustainable way of P nutrition of crops. In this regards phosphate-solubilizing microorganisms (PSM) have been seen as best eco-friendly means for P nutrition of crop. Although, several bacterial (*pseudomonads and bacilli*) and fungal strains (*Aspergilli* and *Penicillium*) have been identified as PSM their performance under *in situ* conditions is not reliable and therefore needs to be improved by using either genetically modified strains or co-inoculation techniques. This review focuses on the diversity of PSM, mechanism of P solubilization, role of various phosphatases, impact of various factors on P solubilization, the present and future scenario of their use and potential for application of this knowledge in managing a sustainable environmental system.

INTRODUCTION

Phosphorus is the most important key element in the nutrition of plants, next to nitrogen (N). It plays an important role in virtually all major metabolic processes in plant including photosynthesis, energy transfer, signal transduction, macromolecular biosynthesis and respiration

(Khan et al.2010) and nitrogen fixation in legumes (Saber et al. 2005). Although P is abundant in soils in both inorganic and organic forms, it is a major limiting factor for plant growth as it is in an unavailable form for root uptake. Inorganic P occurs in soil, mostly in insoluble mineral complexes, some of them appearing after frequent application of chemical fertilizers. These insoluble, precipitated forms cannot be absorbed by plants (Rengel and Marschner 2005). Organic matter is also an important reservoir of immobilized P that accounts for 20–80% of P in soils (Richardson 1994). Only 0.1% of the total P exists in a soluble form available for plant uptake (Zhou et al. 1992) because of its fixation into an unavailable form due to P fixation. The term P fixation is used to describe reactions that remove available phosphate from the soil solution into the soil solid phase (Barber 1995). There are two types of reactions (a) phosphate sorption on the surface of soil minerals and (b) phosphate precipitation by free Al^{3+} and Fe^{3+} in the soil solution (Havlin et al.1999). The soils that exhibit highest P fixation capacity occupy 1,018 million hectares (ha) in the tropics (Sanchez and Logan 1992). It is for this reason that soil P becomes fixed and available P levels have to be supplemented on most agricultural soils by adding chemical P fertilizers, which not only represent a major cost of agricultural production but also impose adverse environmental impacts on overall soil health and degradation of terrestrial, freshwater and marine resources (Tilman et al. 2001). Thus, increased P levels have been identified as a main factor for eutrophication of surface waters that may lead to algal blooms (Schindler et al. 2008). The repeated and injudicious applications of chemical P fertilizers, leads to the loss of soil fertility (Gyaneshwar et al. 2002) by disturbing microbial diversity, and consequently reducing yield of crops. The long-term effect of different sources of phosphate fertilizers on microbial activities includes inhibition of substrate-induced respiration by streptomycin sulphate (fungal activity) and actidione (bacterial activity) and microbial biomass carbon (C) (Bolan et al. 1996). Similarly, the application of triple superphosphate (94 kg/ ha) has shown a substantial reduction in microbial respiration and metabolic quotient (qCO_2) (Chandini and Dennis 2002).

Moreover the efficiency of applied P fertilizers in chemical form rarely exceeds 30% due to its fixation, either in the form of iron/ aluminium phosphate in acidic soils (Norrish and Rosser 1983) or in the form of calcium phosphate in neutral to alkaline soils (Lindsay et al. 1989). It has been suggested that the accumulated P in agricultural soils would be sufficient to sustain maximum crop yields worldwide for about 100 years if it were available (Khan et al. 2009a, b, c). A major characteristic of P biogeochemistry is that only 1% of the total soil P (400–4,000 kg P/ ha in the top 30 cm) is incorporated into living plant biomass during each growing season (10–30 kg P/ha), reflecting its low availability for plant uptake (Blake et al. 2000; Quiquampoix and Mousain 2005). Furthermore P is a finite resource and based on its current rate of use, it has been estimated that the worlds known reserves of high quality rock P may be depleted within the current century (Cordell et al. 2009). Beyond this time the production of P based fertilizers will require the processing of lower grade rock at significantly higher cost (Isherwood 2000). The realization of all these potential problems associated with chemical P fertilizers together with the enormous cost involved in their manufacture, has led to the search for environmental compatible and economically feasible alternative strategies for improving crop production in low or P-deficient soils (Zaidi et al. 2009). The use of microbial inoculants (biofertilisers) possesing P-solubilizing activities in agricultural soils is considered as an environmental-friendly alternative to further applications of chemical based P fertilizers.

Constraints in Using Phosphate Fertilizers

There is global concern about the energy and costs involved in mining the rock phosphate and its transport from manufacturing sites to farm crop fields. Mining phosphate minerals and spreading P fertilizers over the landscape is neither eco-friendly, economically feasible nor it is sustainable and it poses following constraints (i) emission of the fluorine as the highly volatile and poisonous HF gas, (ii) disposal of gypsum and (iii) accumulation of Cd and other heavy metals in soil and possibly crops as a result of repetitive use of P fertilizers.At present mining rate (about 7,100 million tones/annum), reserve will be depleted in about 500–600 years. In India, deposits of sufficiently enriched phosphatic rocks are limited and hence it imports 2 million tons of rock phosphate annually. About 98% of cropland in India is deficient in available forms of soil phosphorus and only 1-9% has high phosphorus status. Intensive cropping pattern during this green and white revolution has also resulted in widespread deficiency of phosphorus. Although various amendments are available for management of P in different soil, all are costlier and practically difficult. Thus, even if the total soil P is high and also if P fertilizers are applied regularly, pH dependent chemical fixation determines the quantity of available P. The holistic P management involves a series of strategies involving manipulation of soil and rhizosphere processes, development of P efficient crops and improving P recycling efficiency. Microbial mediated P management is an ecofriedly and cost effective approach for sustainable development of agricultural crop.

Microorganisms are an integral component of the soil P cycle and are important for the transfer of P between different pools of soil P. Phosphate Solubilzing Microorganisms (PSM) through various mechanisms of solubilization and mineralisation are able to convert inorganic and organic soil P respectively (Khan et al. 2009a) into the bioavailable form facilitating uptake by plant roots. It is important to determine the actual mechanism of P solubilisation by PSM for optimal utilization of these microorganisms in varied field conditions. Hence it is imperative to better understand the plant-soil-microbial P cycle with the aim of reducing reliance on chemical P fertilizers. This has led to increased interest in the harnessing of microorganisms to support P cycling in agroecosystems.

Occurrence and isolation Of PSM

Solubilization of insoluble P by microorganisms was reported by Pikovskaya (1948). During the last two decades knowledge on phosphate solubilizing microorganisms increased significantly (Richardson 2001; Rodriguez and Fraga 1999). Several strains of bacterial and fungal species have been described and investigated in detail for their phosphate-solubilizing capabilities (Glick 1995; He et al. 1997). Typically such microorganisms have been isolated using cultural procedures with species of *Pseudomonas* and *Bacillus* bacteria (Illmer and Schinner 1992) and *Aspergillus* and*Penicillium* fungi being predominant (Wakelin et al. 2004). These organisms are ubiquitous but vary in density and mineral phosphate solubilizing (mps) ability from soil to soil or from one production system to another. In soil P solubilizing bacteria constitute 1-50% and fungi 0.1-0.5% of the total respective population. They are generally isolated from rhizosphere and nonrhizosphere soils, rhizoplane, phyllosphere, and rock P deposit area soil and even from stressed soils using serial plate dilution method or by enrichment culture technique (Zaidi et al. 2009).

The concentration of iron ore, temperature, and C and N sources greatly influence the P-solubilizing potentials of these microbes. Among the various nutrients used by these microorganisms, ammonium salts has been found to be the best N source followed by asparagine, sodium nitrate, potassium nitrate, urea and calcium nitrate (Ahuja et al. 2007). Since 1948, when Pikovskaya suggested that microbes could dissolve non-readily available forms of soil P and play an important role in providing P to plants, numerous methods and media, such as Pikovskaya (Pikovskaya 1948), bromophenol blue dye method (Gupta et al. 1994) and National Botanical Research Institute P (NBRIP) medium (Nautiyal 1999) have been proposed. The source of insoluble phosphate in the culture media to isolate PSM is a major issue of controversy regarding the isolation of PSM in true sense. Commonly used selection factor for this trait, tricalcium phosphate (TCP), is relatively weak and unreliable as a universal selection factor for isolating and testing phosphate-solubilizing microorganisms (PSM) for enhancing plant growth. The use of TCP usually yields many (up to several thousand per study) isolates of "supposed" PSM. When these isolates are further tested for direct contribution of phosphorus to the plants, only a very few are true PSM. Other compounds are also tested, but on a very small scale. These phosphates, mainly iron/aluminium phosphate and several calcium phosphates are even less soluble than TCP in water. Because soils greatly vary in pH and several chemical properties, it appears that there is no metal-Phosphate compound that can serve as the universal selection factor for PSM. Here multiple sources of insoluble phosphate are recommended. The selection of the metal-Phosphate candidates for potential PSM will depend on the type of soil (alkaline, acidic, or organic-rich) where the PSM will be used. Adding calcium phosphate compounds (including rock phosphates) for alkaline soils, iron/aluminium phosphate compounds for acidic soils, and phytates for soils rich in organic P is suggested Bashan et al. 2013a, b.

Both bacterial and fungal strains exhibiting P solubilizing activity are detected by the formation of clear halo (a sign of solubilization) around their colonies. Production of a halo on a solid agar medium should not be considered the sole test for P solubilization. When colonies grow without a halo after several replacements of the medium, an additional test in liquid media to assay P dissolution should be performed and the few isolates that are obtained after such rigorous selection should be further tested for abundant production of organic acids and the isolates complying with these criteria should be tested on a model plant as the ultimate test for potential P solubilization (Bashan et al. 2013a). The viable microbial preparations possessing P-solubilizing activity are generally termed as microphos (Zaidi et al. 2009). The phosphate-solubilizing microbes showing greater solubilization (both qualitatively and quantitatively) of insoluble P under in vitro conditions are selected for field trials prior to production in bulk for ultimate transmission as a biofertiliser.

Once a potential isolate is identified, it must be further tested for direct contribution to P plant nutrition and not necessarily to general growth promotion, as commonly done because promotion of growth, even by PSB, can be the outcome of other mechanisms. (Bashan et al. 2013a) and abilty to solubilise P is not necessarily correlated with the ability to promote plant growth (Collavino et al. 2010).

The production of biofertilizer and its acceptance by farming communities are closely linked. For uptake by farmers, quality management is essential and must be performed consistently in order to supply reliable and contaminant-free bio products. As far as *in vitro* field trials are concerned the establishment and performance of these PSM inoculate developed in laboratory is largely hampered by environmental variables including salinity, pH, moisture, temperature and climatic conditions of the soil. Moreover it is also known that inocula developed from a particular soil fail to function as effectively in soils having different properties (Rodriguez and Fraga 1999). Hence there is a need to study PSM activity in correlation with these factors before PSM application as a biofertiliser. Protocol for isolation and effective inoculants development of PSM based biofertiliser has been shown in Figure 1.

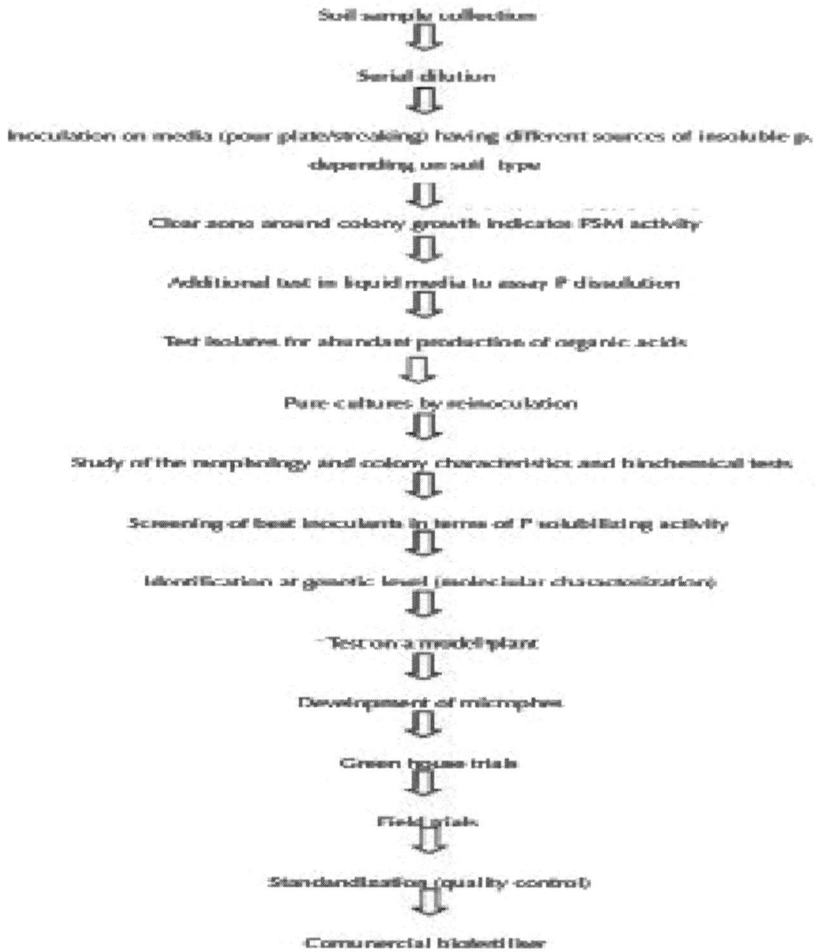

Figure 1: Protocol for isolation and development of effective inoculants of PSM based biofertiliser.

Biodiversity of P Solubilizers

A substantial number of microbial species exhibit P solubilization capacity; these include bacteria, fungi, actinomycetes and even algae. In addition to *Pseudomonas* and *Bacillus*, other bacteria reported as P-solubilizers include *Rhodococcus, Arthrobacter, Serratia, Chryseobacterium, Gordonia, Phyllobacterium, Delftia sp.* (Wani et al. 2005; Chen et al. 2006), *Azotobacter* (Kumar et al. 2001), *Xanthomonas* (De Freitas et al. 1997), *Enterobacter, Pantoea, and Klebsiella* (Chung et al. 2005), *Vibrio proteolyticus, Xanthobacter agilis* (Vazquez et al. 2000). Furthermore, symbiotic nitrogenous *rhizobia,* which fix atmospheric nitrogen into ammonia and export the fixed nitrogen to the host plants, have also shown PS activity (Zaidi et al. 2009) For instance, *Rhizobium leguminosarum bv. Trifolii* (Abril et al. 2007), and *Rhizobium* species nodulating *Crotalaria* species (Sridevi et al. 2007) improved plant P-nutrition by mobilizing inorganic and organic P. Various PS bacteria have also been isolated from stressed environments for example the halophilic bacteria*Kushneria sinocarni* isolated from the sediment of Daqiao saltern on the eastern coast of China, which may be useful in salt affected agricultural soils (Zhu et al. 2011).

In soil, P-solubilizing fungi constitute about 0.1–0.5% of total fungal populations (Kucey 1983). Moreover, P-solubilizing fungi do not lose the P dissolving activity upon repeated sub culturing under laboratory conditions as occurs with the P-solubilizing bacteria (Sperber 1958a, b; Kucey1983). Moreover, fungi in soils are able to traverse long distances more easily than bacteria and hence, may be more important to P solubilization in soils (Kucey 1983). Generally, the P-solubilizing fungi produce more acids than bacteria and consequently exhibit greater P-solubilizing activity (Venkateswarlu et al. 1984). Among filamentous fungi that solubilize phosphate, the genera *Aspergillus* and *Penicillium* (Fenice et al. 2000; Khan and Khan 2002; Reyes et al. 1999,2002) are the most representative although strains of *Trichoderma* (Altomare et al. 1999) and*Rhizoctonia solani* (Jacobs et al. 2002) have also been reported as P solubilizers. A nematofungus*Arthrobotrys oligospora also* has the ability to solubilize phosphate *in vivo* as well as *in vitro* (Duponnois et al. 2006). Among the yeasts, only a few studies have been conducted to assess their ability to solubilize phosphate these include *Yarrowia lipolytica* (Vassilev et al. 2001),*Schizosaccharomyces pombe* and *Pichiafermentans*. As more studies are conducted, a wider diversity of phosphate-solubilizing filamentous fungi are expected to be described. Of those identified, many are commonly found in agricultural soils such as *Penicillium sp., Mucor sp.* and*Aspergillus sp.* which has been shown to increase plant growth by 5–20% after inoculation (Gunes et al. 2009).

The P-solubilizing ability of actinomycetes has attracted interest in recent years because this group of soil organisms is not only capable of surviving in extreme environments (e.g. drought, fire.) but also possess other potential benefits (e.g. production of antibiotics and phytohormone-like compounds) that could simultaneously benefit plant growth (Fabre et al. 1988; Hamdali et al.2008 a, b). A study by Hamdali et al. (2008a) has indicated that approximately 20% of actinomycetes can solubilize P, including those in the common genera *Streptomyces* and*Micromonospora*. A partial list of PSM including various groups is given in Table 1.

Table 1: Biodiversity of PSM

Bacteria	Alcaligenes sp., Aerobactor aerogenes, Achromobacter sp., Actinomadura oligospora, Agrobacterium sp., Azospirillum brasilense, Bacillus sp., Bacillus circulans, B.cereus, B.fusiformis, B. pumils, B. megaterium, B. mycoides, B. polymyxa, B. coagulans B,.chitinolyticus, B. subtilis, Bradyrhizobium sp., Brevibacterium sp., Citrobacter sp., Pseudomonas sp., P putida, P. striata, P. fluorescens, P. calcis, Flavobacterium sp., Nitrosomonas sp., Erwinia sp., Micrococcus sp., Escherichia intermedia, Enterobacter asburiae, Serratia phosphoticum, Nitrobacter sp., Thiobacillus ferroxidans, T. thioxidans, Rhizobium meliloti, Xanthomonas sp.
Fungi	Aspergillus awamori, A. niger, A. tereus, A. flavus, A. nidulans, A. foetidus, A. wentii. Fusarium oxysporum, Alternaria teneius, Achrothcium sp. Penicillium digitatum, P lilacinium, P balaji, P. funicolosum, Cephalosporium sp. Cladosprium sp., Curvularia lunata, Cunnighamella, Candida sp., Chaetomium globosum, Humicola inslens, Humicola lanuginosa, Helminthosporium sp., Paecilomyces fusisporous, Pythium sp., Phoma sp., Populospora mytilina, Myrothecium roridum, Morteirella sp., Micromonospora sp., Oideodendron sp., Rhizoctonia solani, Rhizopus sp., Mucor sp., Trichoderma viridae, Torula thermophila, Schwanniomyces occidentalis, Sclerotium rolfsii.
Actinomycetes	Actinomyces,, Streptomyces.
Cyanobacteria	Anabena sp., Calothrix braunii, Nostoc sp., Scytonema sp.,
VAM	Glomus fasciculatum.

Sharma *et al.*

In addition to bacteria, fungi and actinomycetes, algae such as cyanobacteria and mycorrhiza have also been reported to show P solubilization activity.The interactive effects of arbuscular mycorrhizal fungi (AMF) and rhizobacteria on the growth and nutrients uptake of Sorghum bicolor were studied in acid and low availability phosphate soil. The microbial inocula consisted of the AMFs *Glomus manihotis* and *Entrophospora colombiana,* PSB *Pseudomonas* sp., results indicated that the interaction of AMF and the selected rhizobacteria has a potential to be developed as biofertilizers in acid soil. The potential of dual inoculation with AMF and rhizobacteria needs to be further evaluated under different crop and agroclimatic conditions, particularly in the field (Widada et al. 2007).

Hence the studies have shown that the diversity of the PSM's is highly varied in different ecological niches and there is ample scope to identify many new potent isolates from varied environments in coming times.

Mechanism of P-Solubilization by PSM

In a review of P chemistry in soils, Sims and Pierzynski (2005) identified the major processes of the soil P cycle that affect soil solution P concentrations as(1) dissolution–precipitation (mineral equilibria), (2) sorption–desorption (interactions between P in solution and soil solid surfaces), and (3) mineralization–immobilization (biologically mediated conversions of P between inorganic and organic forms).

The main P solubilization mechanisms employed by soil microorganisms include: (1) release of complexing or mineral dissolving compounds e.g. organic acid anions, siderophores, protons, hydroxyl ions, CO_2, (2) liberation of extracellular enzymes (biochemical P mineralization) and (3) the release of P during substrate degradation (biological P mineralization) (McGill and Cole 1981). Therefore, microorganisms play an important role in all three major components of the soil P cycle (i.e. dissolution–precipitation, sorption–desorption, and mineralization–immobilization). Additionally these microorganisms in the presence of labile C serve as a sink for P, by rapidly immobilizing it even in low P soils; therefore PSM become a source of P to plants upon its release from their cells. Release of P immobilized by PSM primarily occurs when cells die due to changes in environmental conditions, starvation or predation. Environmental changes, such as drying–rewetting or freezing–thawing, can result in so-called flush-events, a sudden increase in available P in the solution due to an unusually high proportion of microbial cell lysis (Butterly et al. 2009). (Grierson et al. 1998) found that about 30–45% of microbial P (0.8–1 mg kg^{-1}) was released in a sandy spodosol in an initial flush after drying–rewetting cycles within the first 24 hour.

Inorganic Psolubilization

By P-solubilizing microorganisms occurs mainly by organic acid production (Table 2), either by: (i) lowering the pH, or (ii) by enhancing chelation of the cations bound to P (iii) by competing with P for adsorption sites on the soil (iv) by forming soluble complexes with metal ions associated with insoluble P (Ca, Al, Fe) and thus P is released. The lowering in pH of the medium suggests the release of organic acids by the P-solubilizing microorganisms (Whitelaw 2000; Maliha et al. 2004) via the direct oxidation pathway that occurs on the outer face of the cytoplasmic membrane (Zaidi et al. 2009). These acids are the product of the microbial metabolism, mostly by oxidative respiration or by fermentation of organic carbon sources (e.g., glucose) (Atlas and Bartha 1997; Trolove et al. 2003) or such organic acids can either directly dissolve the mineral P as a result of anion exchange of phosphate by acid anion or can chelate Fe, Al and Ca ions associated with P (Omar 1998).

Table 2: Important PSM, their ecological niches and organic acids produced

Organism	Ecological niche	Predominant acids produced	Reference
PSB	Soil and phosphate bearing rocks	ND(not determined)	Pikovskaya 1948
PSB	Bulk and rhizospheric soil	ND(not determined)	Gerretson 1948
Escherichia freundii	Soil	Lactic	Sperber 1958a, b

Aspergillus niger, Penicillium sp.	Soil	Citric, glycolic, succinic, gluconic, oxalic, lactic	Sperber 1958a, b
Bacillus megaterium, Pseudomonas sp., Bacillus subtilus	Rhizospheric soil	Lactic, malic	Taha et al. 1969
Arthrobacter sp., Bascillus sp., Bacillus firmus B-7650	Wheat and cowpea rhizosphere	Lactic, citric	Bajpai and Sundara Rao 1971
Aspergillus sp., Penicillium sp., Chaetomiumnigricolor	Lateritic soil	Oxalic, Succinic, Citric, 2-ketogluconic	Banik and Dey 1983
A. japonicus, A. foetidus	Indian Rock phosphate	Oxalic, citric, gluconic succinic, tartaric acid	Singal et al. 1994
P. radicum	rhizosphere of wheat roots,	Gluconic	Whitelaw et al. 1999
Enterobacteragglomerans	Wheat rhizosphere	Oxalic, citric	Kim et al. 1997
Bacillus amyloliquefaciens, B. licheniformis, B. atrophaeus, Penibacillus macerans, Vibrio proteolyticus, xanthobacter agilis, Enterobacter aerogenes, E. taylorae, E. asburiae, Kluyvera cryocrescens, Pseudomonas aerogenes, Chryseomonas Luteola	Mangrove ecosystem	Lactic, itaconic, isovaleric, isobutyric, acetic	Vazquez et al. 2000
Penicillium rugulosum	Venezuelan phosphate rocks	Citric, gluconic acid	Reyes et al. 2001
Enterobacter intermedium	Grass rhizosphere	2-ketogluconic	Hwangbo et al. 2003
Aspergillus flavus, A. niger, Penicilliumcanescens	stored wheat grains	Oxalic, citric, gluconic succinic	Maliha et al. 2004
P.fluorescens	Root fragments and rhizosphere of oil palm trees	Citric, malic, tartaric, gluconic	Fankem et al. 2006
Aspergillus niger	Tropical and subtropical soil	Gluconic, oxalic	Chuang et al. 2007

P.trivialis	rhizosphere of Hippophaer hamnoides growing in the cold deserts of Lahaul and Spiti in the trans-Himalayas	Lactic, formic	Vyas and Gulati 2009
B.pumilus var.2;B. subtilisvar.2;Actinomadura oligospora; Citrobacter sp.	Giant Cardon cactus (P.pringlei) growing in ancient lava	Gluconic, Propionic, Isovaleric, Heptonic, Caproic, Isocaproic, Formic, Valeric, Succinic, Oxalic, Oxalacetic, Malonic.	Puente et al. 2004a
B.pumilus CHOO8A; B.fusiformis	Cholla cactus(Opuntia Cholla)		Puente et al. 2004a
Bacillus sp. SENDO 6 and	Giant Cardon cactus (P.pringlei)	Gluconic, Propionic, Isovaleric, Formic, Succinic, Lactic.	Puente et al. 2009a, b
Pseudomonas putida M5TSA, Enterobacter sakazakii M2PFe, and Bacillus megaterium M1PCa	Wild cactus Mammillaria fraileana		Lopez et al. 2011

Sharma *et al.*

The monovalent anion phosphate $H_2PO_4^-$ is a major soluble form of inorganic phosphate, which usually occurs at lower pH. However as the pH of the soil environment increases the divalent and trivalent forms of Pi (HPO_4^{-2} and HPO_4^{-3} respectively) occur. Thus, the synthesis and discharge of organic acid by the PSM strains into the surrounding environment acidify the cells and their surrounding environment that ultimately lead to the release of P ions from the P mineral by H^+substitution for the cation bound to phosphate (Goldstein 1994). The prominent acids released by PSM in the solubilization of insoluble P are gluconic acid (Di-Simine et al. 1998; Bar-Yosef et al.1999), oxalic acid, citricacid (Kim et al. 1997), lactic acid, tartaric acid, aspartic acid (Venkateswarlu et al. 1984). Evidence from an abiotic study using HCl and gluconic acid to solubilize P also indicated that chelation of Al^{3+} by gluconic acid may have been a factor in the solubilization of colloidal Al phosphate (Whitelaw et al. 1999). Organic acids produced by P-solubilizing microorganisms can be detected by high performance liquid chromatography and enzymatic methods (Parks et al. 1990; Whitelaw 2000). However, acidification does not seem to be the only mechanism of solubilization, as the ability to reduce the pH in some cases did not correlate with the ability to solubilize mineral P (SubbaRao 1982). Altomare et al. (1999) investigated the capability of the plant-growth promoting and biocontrol fungus *T. harzianum T-22*to solubilize in vitro insoluble minerals including rock phosphate. Organic acids were not detected in the culture filtrates and hence, the authors concluded that acidification was probably not the major mechanism of solubilization as the pH never fell below 5. The phosphate solubilizing activity was attributed both to chelation and to reduction processes. Although, organic acid has been suggested as the principal mechanism of P solubilization, the solubilization of insoluble P by inorganic acid (e.g. HCl) has also been reported, although HCl was able to solubilize less P from hydroxyapatite than citric acid or oxalic acid at same pH (Kim et al. 1997). Bacteria of the genera*Nitrosomonas* and *Thiobacillus* species can also dissolve phosphate compounds by producing nitric and sulphuric acids (Azam and Memon 1996).

According to the sink theory, P-solubilizing organisms remove and assimilate P from the liquid and hence, activate the indirect dissolution of calcium phosphate compounds by consistent removal of P from liquid culture medium. For instance, the P content in the biomass of*Pseudomonas* sp. and *P. aurantiogriseum* were similar to those observed in non-P-solubilizing microorganisms (Illmer et al. 1995) which can be explained by the fact that the P content in biomass of organisms is consistently correlated with the decomposition of P containing organic substrates (Dighton and Boddy 1989).

The other mechanism is the production of H2S, which react with ferric phosphate to yield ferrous sulphate with concomitant release of phosphate (Swaby and Sperber 1958).

(Rudolph It has been suggested that MPS activity occurs as a consequence of microbial sulphur oxidation1922), nitrate production and CO_2 formation These processes result in the formation of inorganic acids like sulphuric acid (Sperber 1958a). However, their effectiveness has been less accepted than the concept of involvement of organic acids in solubilization (Kim et al. 1997).

H^+ Excretion Originating from NH_4^+ Assimilation

H^+ excretion originating from NH_4^+ assimilation as proposed by Parks et al. (1990) could be the alternative mechanisms of P solubilization. An HPLC analysis of the culture solution of*Pseudomonas* sp., in contrast to the expectation, did not detect any organic acid while solubilization occurred (Illmer and Schinner 1995). They also reported that the most probable reason for solubilization without acid production is the release of protons accompanying respiration or NH_4^+ assimilation. Krishnaraj et al. (1998) have proposed a model highlighting the importance of protons that are pumped out of the cell to be the major factor responsible for P solubilization. Here direct role of organic or inorganic acids has been ruled out. For some microorganisms, NH_4^+driven proton release seems to be the sole mechanism to promote P solubilization. Asea et al. (1988) tested two fungi, *Penicillium bilaii* and *Penicillium fuscum*, for their ability to solubilize phosphate rock in the presence of NH_4^+ or without N addition, and showed that only *P. bilaii*maintained the ability to decrease the pH and mobilize P when no N was supplied. In a study of*Pseudomonas fluorescens*, the form of C supply (e.g. glucose versus fructose) rather than N supply (e.g. NH_4^+ versus NO_3^-) had the greatest effect on proton release (Park et al. 2009). Further, the involvement of the H^+ pump mechanism in the solubilization of small amounts of P in*Penicillium rugulosum* is reported (Reyes et al. 1999). Acidification of the rhizosphere of cactus seedlings (giant cardon, *Pachycereus pringlei*) after inoculation with the plant growth-promoting bacterium *Azospirillum brasilense*, in the presence or absence of ammonium and nitrate, was studied and it was assumed that the effect of inoculation with this PGPB on plant growth, combined with nitrogen nutrition, might be affecting one or more of the metabolic pathways of the plant which increases proton efflux from roots and liberation of organic acid, leading to rhizosphere acidification (Carrillo et al. 2002).This indicates that for different species, different mechanisms are responsible for proton release, only partly depending on the presence of NH_4^+.

Goldstein (1995) suggested that extracellular oxidation via direct oxidation pathway may play an essential role in soils where calcium phosphates provide a significant pool of unavailable mineral phosphorus. This has been confirmed by some researchers (Song et al. 2008) by biochemical analysis of lowering of pH in insoluble P solubilization by *Burkholderia cepacia* DA23.

Organic P Solubilization

Organic P solubilization is also called mineralization of organic phosphorus. Mineralization of soil organic P (Po) plays an imperative role in phosphorus cycling of a farming system. Organic P may constitute 4–90% of the total soil P (Khan et al. 2009b). Such P can be released from organic compounds in soil by enzymes:

Non-specific Acid Phosphatases (NSAPs)

Non-specific acid phosphatases (NSAPs), which dephosphorylate phospho-ester or phosphoanhydride bonds of organic matter. Among the variety of phosphatase enzyme classes released by PSM, phosphomonoesterases (often just called phosphatases) are the most abundant and best studied (Nannipieri et al. 2011). Depending on their pH optima, these enzymes are divided into acid and alkaline phosphomonoesterases and both can be produced by PSM depending upon the external conditions (Kim et al. 1998; Jorquera et al. 2008). Typically, acid phosphatases predominate in acid soils, whereas alkaline phosphatases are more abundant in neutral and alkaline soils (Eivazi and Tabatabai 1977; Juma and Tabatabai 1977, 1998; Renella et al. 2006). Although plant roots can produce acid phosphatases they rarely produce large quantities of alkaline phosphatases, suggesting that this is a potential niche for PSM (Juma and Tabatabai 1998; Criquet et al. 2004). It is also difficult to differentiate between root- and PSM-produced phosphatases (Richardson et al. 2009a, b) but some evidence suggests that phosphatases of microbial origin possess a greater affinity for Po compounds than those derived from plant roots (Tarafdar et al. 2001). The relationship between PSM introduced into soil, phosphatase activity and the subsequent mineralization of Po still remains poorly understood (Chen et al. 2003)

Phytases

Phytases, which specifically cause release of P from phytate degradation. In its basic form, phytate is the primary source of inositol and the major stored form of P in plant seeds and pollen, and is a major component of organic P in soil (Richardson, 1994). Although the ability of plants to obtain P directly from phytate is very limited, yet the growth and P-nutrition of Arabidopsis plants supplied with phytate was significantly improved when they were genetically transformed with the phytase gene (phyA) derived from *Aspergillus niger* (Richardson et al. 2001). This led to an increase in P-nutrition to such an extent that the growth and P-content of the plant was equivalent to control plants supplied with inorganic P. Hence microorganisms are in fact a key driver in regulating the mineralization of phytate in soil and their presence within the rhizosphere may compensate for a plants inability to otherwise acquire P directly from phytate (Richardson and Simpson 2011).

Phosphonatases and C–P lyases

Phosphonatases and C–P lyases, that cleave the C–P bond of organophosphonates (Rodriguez et al. 2006).

It is therefore clear that P solubilization by PSMs has been a subject of analysis and research for a long time and still the research seems to be in its infancy. It occurs through different mechanisms and there is considerable variation amongst the organisms in this respect. Each organism can act in one or more than one way to bring about the solubilization of insoluble P. Though it is difficult to pin point a single mechanism, production of organic acids and consequent pH reduction appears to be of great importance. Different mechanisms involved in the solubilization and mineralization of insoluble P by naturally-occurring microbial communities of soils are briefly illustrated in Figure 2.

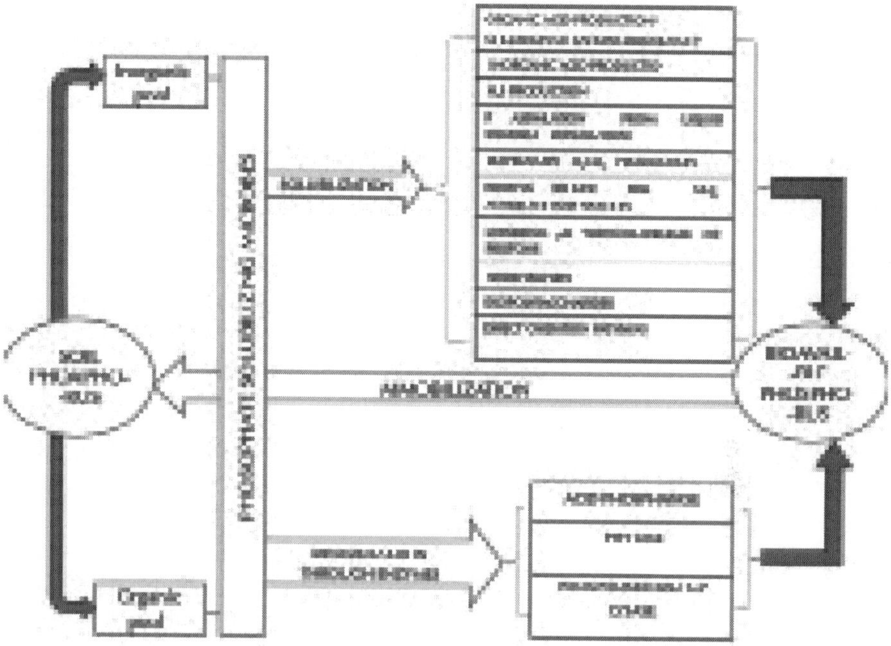

Figure 2: Schematic representation of mechanism of soil P solubilization/ mineralization and immobilization by PSM.

Role of Siderophores in P Solubilization

Siderophores are complexing agents that have a high affinity for iron and are produced by almost all microorganisms in response to iron deficiency. Thus siderophores act as solubilizing agents for iron from minerals or organic compounds under conditions of iron limitation. There are approximately 500 known siderophores, with the majority of them being used by a wide range of microorganisms and plants and some of them being exclusively used by the microbial species and strains that produce them (Crowley 2007). Studies have reported the release of siderophores from PSM (Vassilev et al. 2006; Caballero-Mellado et al. 2007; Hamdali et al. 2008a); however, siderophore production has not been widely implicated as a P-solubilization mechanism. Considering the dominance of mineral dissolution over ligand exchange by organic acid anions as a P-solubilizing mechanism (Parker et al. 2005), the potential role of siderophores in enhancing P availability should be obvious.

Role of EPS in P Solubilization

Recently the role of polysaccharides in the microbial mediated solubilization of P was assessed by Yi et al. (2008). Microbial exopolysaccharides (EPSs) are polymers consisting mainly of carbohydrates excreted by some bacteria and fungi onto the outside of their cell walls. Their composition and structures are very varied; they may be homo- or heteropolysaccharides and may also contain a number of different organic and inorganic substituents (Sutherland 2001). Four bacterial strains of *Enterobacter sp. (EnHy-401), Arthrobacter sp. (ArHy-505), Azotobacter sp. (AzHy-510) and Enterobacter sp. (EnHy-402),* possessing the ability to solubilize TCP (tri calcium phosphate), were used to assess the role of exopolysaccharide (EPS) in the solubilization of P by Yi et al. 2008. These Phosphate Solubilizing bacteria produced a significant amount of EPS and demonstrated a strong ability for P-solubilization. However further studies are necessary to understand the relationship between EPS production and phosphate solubilization.

Plant Growth Promoting Attributes of PSM

Besides making soluble P accessible for uptake by plants, there have been a number of reports on plant growth promotion by these microorganisms (Gaur and Ostwal 1972). This Santibiotics, or siderophores. Various PSM preparations have been shown to promote the growth of many crops (Table 3). Endophytic Bacteria isolated from rhizoplane of cacti growing in bare lava rocks, not only significantly mobilized Phosphate and other minerals (Puente et al. 2004a, 2009a) but also promoted growth of wild cactus species (Puente et al. 2004b, b). The mechanisms involved in plant growth promotion by PSM are outlined in Figure 3.

Table 3: Plant growth promotion by PSM (Patil et al 2002)

PSM Bioinoculant	Crop benefited
B. firlmus NCIM 2636	Paddy in acid soils

G. faciculatum	Banana
B. megaterium+G. faciculatum	Banana
Phosphobacterium	Sword bean variety SBS 1
P. Striata	Soybean in sandy alluvial soil
P. Striata	Chick pea

Sharma *et al.*

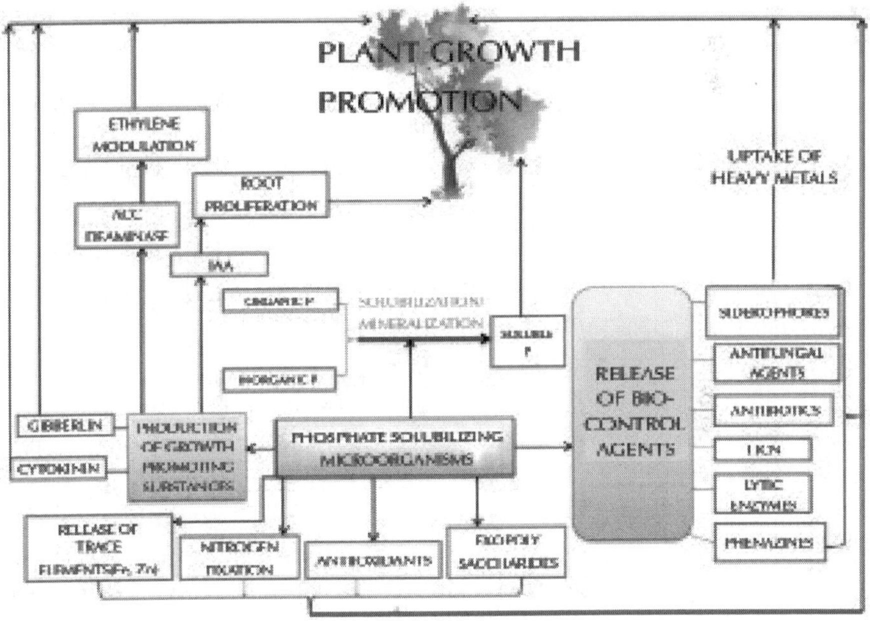

Figure 3: Possible Mechanisms involved in plant growth promotion by PSM.

Genetic Engineering of PSM

Although knowledge of the genetics of phosphate solubilization is still scanty, and the studies at the molecular level in order to understand how precisely the PSM brings out the solubilization of insoluble P are inconclusive (Rodriguez et al. 2006). However, some genes involved in mineral and organic phosphate solubilization have been isolated and characterized. Initial achievements in the manipulation of these genes through genetic engineering and molecular biotechnology followed by their expression in selected rhizobacterial strains open a promising perspective for obtaining PSM strains with enhanced phosphate solubilizing capacity, and thus, a more effective use of these microbes as agricultural inoculants. The initial achievement in cloning of gene involved in P solubilization from the Gram negative bacteria *Erwinia herbicola* was achieved by Goldstein and Liu (1987). Similarly the napA phosphatase gene from the soil bacterium *Morganella morganii* was transferred to *Burkholderia cepacia IS-16*, a strain used as a biofertilizer, using the broad-host range vector pRK293 (Fraga et al. 2001). An increase in extracellular phosphatase activity of the recombinant strain was achieved.

Introduction or over-expression of genes involved in soil phosphate solubilization (both organic and inorganic) in natural rhizosphere bacteria is a very attractive approach for improving the capacity of microorganisms to work as inoculants. Insertion of phosphate-solubilizing genes into microorganisms that do not have this capability may avoid the current need of mixing two populations of bacteria, when used as inoculants (nitrogen fixers and phosphate-solubilizers (Bashan et al. 2000). There are several advantages of developing genetically-modified PSM over transgenic plants for improving plant performance: (1) With current technologies, it is far easier to modify a bacterium than complex higher organisms, (2) Several plant growth-promoting traits can be combined in a single organism, and (3) Instead of engineering crop by crop, a single, engineered inoculant can be used for several crops, especially when using a non-specifc genus like Azospirillum (Rodriguez et al. 2006). Some barriers should be overcome first to achieve successful gene insertions using this approach, such as the dissimilarity of metabolic machinery and different regulating mechanism between the donor and recipient strains. Despite the diffculties, significant progress has been made in obtaining genetically engineered microorganisms for agricultural use (Armarger 2002). Overall, further studies on this aspect of PSM will provide crucial information in future for better use of these PSM in varied environmental conditions.

Current Trends

Phosphorus is an important limiting factor in agriculture production, and considering the negative effects of chemical P fertilizers, microbial intervention of PSM seems to be an effective way to solve the phosphorus availability in soil. However P-solubilization in soil is much more difficult to study than solubilization of P in broth culture. The crops respond differently to the inoculation of PSMs and are dependent on several factors such as the soil temperature, moisture, pH, salinity, and source of insoluble P, method of inoculation, the energy sources and the strain of microorganism used. Hence study of PSM activity in correlation with these factors has to be done extensively before PSM can be used as a biofertiliser with promising results. The successful implementation of this approach has already been demonstrated in the fields by various workers, to a limited extent. However the large scale use of this technology would benefit from additional studies, particularly those directed towards understanding how the interaction between soil and microbial system might be facilitated.

The organisms involved in phosphorus (P) cycling in soils are highly varied, and microorganisms probably play the most important role. However, more than 99% of soil microorganisms have not been cultured successfully (Torsvik and Ovreas 2002). Therefore, culture-independent methods are required to study the function and ecology of microbes involved in P cycling in soils. Molecular approaches for such culture-independent methods have been developed in the recent past. The molecular techniques based on nucleic acid composition like LMW RNA profiling and PCR based techniques, are excellent tools for this purpose, as they are precise, reproducible and not dependent on culture media composition or growth phase of microorganisms (Peix et al. 2007).

Molecular-based techniques also provide new opportunity to detect the presence and abundance of specific microorganisms or to quantify the expression of target genes directly in soil or in the rhizosphere with high levels of sensitivity. For example, specific primers based on conserved regions have been described for various microorganisms associated with P mobilization, including mycorrhizal fungi, Penicillium sp., and Pseudomonas sp. (Oliveira et al. 2009), as have primers that are directed at traits such as bacterial phytases (Jorquera et al. 2011). Microarrays composed of suites of functional bacterial genes and arrays for phylogenetic analysis of bacterial diversity based on 16S-RNAgene sequences along with next-generation sequencing and soil microbiome analyses; provide further application for assessment of diversity surrounding particular traits or functional groups of microorganisms (Richardson and Simpson 2011). Collectively, these tools provide new opportunities to address key questions in microbial community ecology and to assess the survival and persistence of specific inoculants under different environmental conditions.

Looking at the possible avenues which can open up with exploring these environmental friendly microorganisms, it is necessary to study the composition and dynamics of these microbial populations to reach a better understanding of soil PSM diversity and P uptake by plants.

Future Prospects

Despite their different ecological niches and multiple functional properties, P-solubilizing microorganisms have yet to fulfill their promise as commercial bio-inoculants. Current developments in our understanding of the functional diversity, rhizosphere colonizing ability, mode of actions and judicious application are likely to facilitate their use as reliable components in the management of sustainable agricultural systems. Although significant studies related to PSM and their role in sustainable agriculture have been done over the last few decades, the required technique remains in its infancy. Nevertheless with an awareness of the limitations of existing methods, a reassessment can be expected, so that the use of PSM as potential biofertilisers in different soil conditions becomes a reality.

Enhancement in the use of PSM is one of the newly emerging options for meeting agricultural challenges imposed by the still-growing demand for food. Thus, more than ever, obtaining high yields is the main challenge for agriculture. In addition, in recent years both producers and consumers have increasingly focused on the health and quality of foods, as well as on their organoleptic and nutritional properties. Hence, this biotechnology is also likely to ensure conservation of our environments. However, before PSM can contribute to such benefits, scientists must learn more about them and explore ways and means for their better utilization in the farmers' fields. Future research should focus on managing plant–microbe interactions, particularly with respect to their mode of actions and adaptability to conditions under extreme environments for the benefit of plants. Furthermore, scientists need to address certain issues, like how to improve the efficacy of biofertilizers, what should be an ideal and universal delivery system, how to stabilize these microbes in soil systems, and how nutritional and root exudation aspects could be controlled in order to get maximum benefits from PSM application. Biotechnological and molecular approaches could possibly develop more understanding about PSM mode of actions that could lead to more successful plant–microbe interaction. Efforts should also be directed towards the use of PSM to reduce pesticide applications. In brief, PSM biotechnology provides an excellent opportunity to develop environment-friendly phosphorus biofertilizer to be used as supplements and/or alternatives to chemical fertilizers.

CONCLUSIONS

Phosphorus is a vital element in crop nutrition. Adverse environmental effects of chemical based P fertilsers, depleting resources of high grade Phosphatic rocks and their skyrocketing prices have compelled us to find a sustainable approach for efficient P availability in agriculture to meet the ever increasing global demand of food. Soil microorganisms are involved in a range of processes that affect P transformation and thus influence the subsequent availability of P (as phosphate) to plant roots. In particular, microorganisms can solubilize and mineralize P from inorganic and organic pools of total soil P.The use of efficient PSM (phosphate-solubilizing microorganisms), opens up a new horizon for better crop productivity besides sustaining soil health. However, the viability and sustainability of PSM technology largely depends on the development and distribution of good quality inoculants to farming communities. Therefore, there is a need for extensive and consistent research efforts to identify and characterize more PSM with greater efficiency for their ultimate application under field conditions. Soil Scientists and Microbiologists have a great responsibility to the society to find ways and means as to how soil P could be improved without applying the chemical P fertilizers under different agro-climatic regions of the world.

The promise of exploiting soil microorganisms to increase mobilization of soil P remains. Whether or not this will be achieved through better management of soil microbial communities, by development of more effective microbial inoculants, through the genetic manipulation of specific organisms, or with a combination of these approaches is not known. What is clear though is that soil microorganisms play an important role in the mobilization of soil P and that detailed understanding of their contribution to the cycling of P in soil-plant systems is required for the development of sustainable agriculture and our movement from a green revolution to an evergreen revolution can be accompolished.

AUTHORS' CONTRIBUTIONS

SBS: Collected and reviewed the literature and drafted the manuscript. RZS: Provided guidance and improved the manuscript. MHT: provided guidance. GAT: provided guidance. All authors read and approved the final manuscript.

ACKNOWLEDGEMENTS

Financial support from Department Of Science And Technology, Government of India, New Delhi for the fellowship under the WOS-A (Women Scientist-A) scheme is greatfully acknowledged. We extend our gratitude to Dr. Malcolm J. Hawkesford, Rothamsted Research, Harpenden, UK, for his valuable suggestions in preparing the manuscript. We also acknowledge the support from Dr. Mahesh G. Thakkar, Head of the Department, Department of Earth and Environmental Science, KSKV Kachchh university, Bhuj, India, for his positive support in all our endeavours.

REFERENCES

1. Abril A, Zurdo-Pineiro JL, Peix A, Rivas R, Velazquez E (2007) Solubilization of phosphate by a strain of *Rhizobium leguminosarum* bv. *Trifolii* isolated from Phaseolus vulgaris in

El Chaco Arido soil (Argentina). In: Velazquez E, Rodriguez-Berrueco C (eds) Developments in Plant and Soil Sciences, The Netherlands: Springer. pp 135-138

2. Ahuja A, Ghosh SB, D'Souza SF (2007) Isolation of a starch utilizing, phosphate solubilizing fungus on buffered medium and its characterization. Bioresour Technol 98:3408-3411
3. Altomare C, Norvell WA, Borjkman T, Harman GE (1999) Solubilization of phosphates and micronutrients by the plant growth promoting and biocontrol fungus *Trichoderma harzianum Rifai*1295–22. Appl Environ Microbiol 65:2926-2933
4. Armarger N (2002) Genetically modified bacteria in agriculture. Biochimie 84:1061-1072
5. Asea PEA, Kucey RMN, Stewart JWB (1988) Inorganic phosphate solubilization by two Penicillium species in solution culture and soil. Soil Biol Biochem 20:459-464
6. Atlas R, Bartha R (1997) Microbial ecology. New York: Addison Wesley Longman.
7. Azam F, Memon GH (1996) Soil organisms. In: Bashir E, Bantel R (eds) Soil science, Islamabad: National Book Foundation. pp 200-232
8. Bajpai PD, Sundara Rao WVB (1971) Phosphate solubilizing bacteria III. Soil inoculation with phosphate solubilizing bacteria. Soil Sci Plant Nutr 17:46-53
9. Banik S, Dey BK (1983) Phosphate solubilizing potentiality of the microorganisms capable of utilizing aluminium phosphate as a sole phosphate source. Zentralblatt Microbiology 138:17-23
10. Barber SA (1995) Soil nutrient bioavailability. Wiley, New York: A mechanistic approach.
11. Bar-Yosef B, Rogers RD, Wolfram JH, Richman E (1999) *Pseudomonas cepacia*-mediated rock phosphate solubilization in kaolinite and montmorillonite suspensions. Soil Sci Soc Am J 63:1703-1708
12. Bashan Y, Moreno M, Troyo E (2000) Growth promotion of the seawater-irrigated oil seed halophyte Salicornia bigelovii inoculated with mangrove rhizosphere bacteria and halotolerant*Azospirillum* sp. Biol Fertil Soils 32:265-272

13. Bashan Y, Kamnev AA, de Bashan LE (2013a) A proposal for isolating and testing phosphate-solubilizing bacteria that enhance plant growth. Biol Fertil Soils 49:1-2
14. Bashan Y, Kamnev AA, de Bashan LE (2013b) Tricalcium phosphate is inappropriate as a universal selection factor for isolating and testing phosphate-solubilizing bacteria that enhance plant growth: a proposal for an alternative procedure. Biol Fertil Soils 49:465-479
15. Blake L, Mercik S, Koerschens M, Moskal S, Poulton PR, Goulding KWT, Weigel A, Powlson DS (2000) Phosphorus content in soil, uptake by plants and balance in three European long-term field experiments. Nutr Cycl Agroecosyst 56:263-275
16. Bolan NS, Currie LD, Baskaran S (1996) Assessment of the influence of phosphate fertilizers on the microbial activity of pasture soils. Biol Fertil Soils 21:284-292
17. Butterly CR, Bunemann EK, McNeill AM, Baldock JA, Marschner P (2009) Carbon pulses but not phosphorus pulses are related to decrease in microbial biomass during repeated drying and rewetting of soils. Soil Biol Biochem 41:1406-1416
18. Caballero-Mellado J, Onofre-Lemus J, De los Santos EP, Martinez-Aguilar L (2007) The tomato rhizosphere, an environment rich in nitrogen-fixing Burkholderia species with capabilities of interest for agriculture and bioremediation. Appl Environ Microbiol 73:5308-5319
19. Carrillo AE, Li CY, Bashan Y (2002) Increased acidification in the rhizosphere of cactus seedlings induced by *Azospirillum brasilense*. Naturwissenschaften 89:428-432
20. Chandini TM, Dennis P (2002) Microbial activity, nutrient dynamics and litter decomposition in a Canadian Rocky Mountain pine forest as affected by N and P fertilizers. For Ecol Manage 159:187-201
21. Chen CR, Condron LM, Davis MR, Sherlock RR (2003) Seasonal changes in soil phosphorus and associated microbial properties under adjacent grassland and forest in New Zealand. Forest Ecol Manag 117:539-557
22. Chen YP, Rekha PD, Arun AB, Shen FT, Lai WA, Young CC (2006) Phosphate solubilizing bacteria from subtropical soil and their

tricalcium phosphate solubilizing abilities. Appl Soil Ecol 34:33-41

23. Chuang CC, Kuo YL, Chao CC, Chao WL (2007) Solubilization of inorganic phosphates and plant growth promotion by *Aspergillus niger*. Biol Fertil Soils 43:575-584
24. Chung H, Park M, Madhaiyan M, Seshadri S, Song J, Cho H, Sa T (2005) Isolation and characterization of phosphate solubilizing bacteria from the rhizosphere of crop plants of Korea. Soil Biol Biochem 37:1970-1974
25. Collavino MM, Sansberro PA, Mroginski LA, Aguilar OM (2010) Comparison of in vitro solubilization activity of diverse phosphate-solubilizing bacteria native to acid soil and their ability to promote Phaseolus vulgaris growth. Biol Fertil Soils 46:727-738
26. Cordell D, Drangert JO, White S (2009) The story of phosphorus: global food security and food for thought. Glob Environ Chang 19:292-305
27. Criquet S, Ferre E, Farner EM, Le Petit J (2004) Annual dynamics of phosphatase activities in an evergreen oak litter – influence of biotic and abiotic factors. Soil Biol Biochem 36:1111-1118
28. Crowley DE (2007) Microbial siderophores in the plant rhizosphere. In: Barton LL, Abadia J (eds) Iron nutrition in plants and rhizospheric microorganisms, Dordrecht: Springer. pp 169-198
29. De Freitas JR, Banerjee MR, Germida JJ (1997) Phosphate-solubilizing rhizobacteria enhance the growth and yield but not phosphorus uptake of canola (Brassica napus L.). Biol Fertil Soils 24:358-364
30. Dighton J, Boddy L (1989) Role of fungi in nitrogen, phosphorus and sulfur cycling in temperate forest ecosystems. In: Boddy L, Marchant R, Read D (eds) Nitrogen, phosphorus and sulfur utilization by fungi, Cambridge: Cambridge University Press. pp 269-298
31. Di-Simine CD, Sayer JA, Gadd GM (1998) Solubilization of zinc phosphate by a strain of Pseudomonas fluorescens isolated from a forest soil. Biol Fertil Soils 28:87-94

32. Duponnois R, Kisa M, Plenchette C (2006) Phosphate solubilizing potential of the nematofungus*Arthrobotrys oligospora*. J Plant Nutr Soil Sci 169:280-282
33. Eivazi F, Tabatabai MA (1977) Phosphatases in soils. Soil Biol Biochem 9:167-172
34. Fabre B, Armau E, Etienne G, Legendre F, Tiraby G (1988) A simple screening method for insecticidal substances from actinomycetes. J Antibiot 41:212-219
35. Fankem H, Nwaga D, Deube A, Dieng L, Merbach W, Etoa FX (2006) Occurrence and functioning of phosphate solubilizing microorganisms from oil palm tree (*Elaeis guineensis*) rhizosphere in Cameroon. Afr J Biotechnol 5:2450-2460
36. Fenice M, Seblman L, Federici F, Vassilev N (2000) Application of encapsulated Penicilliumvariabile P16 in solubilization of rock phosphate. Bioresour Technol 73:157-162
37. Fraga R, Rodriguez H, Gonzalez T (2001) Transfer of the gene encoding the Nap A acid phosphatase from *Morganella morganii* to a *Burkholderia cepacia* strain. Acta Biotechnol 21:359-369
38. Gaur AC, Ostwal KP (1972) Influence of phosphate dissolving Bacilli on yield and phosphate uptake of wheat crop. Indian J Exp Biol 10:393-394
39. Gerretson FC (1948) The influence of microorganisms on the phosphorus uptake by plant. Plant Soil 1:51-81
40. Glick BR (1995) The enhancement of plant growth by free living bacteria. Can J Microbiol 41:109-117
41. Goldstein AH (1994) Involvement of the quinoprotein glucose dehydrohenase in the solubilization of exogenous phosphates by Gram-negative bacteria. In: Torriani-Gorini A, Yagiland E, Silver S (eds) Phosphate in microorganisms: Cellular and molecular biology, Washington (DC): ASM Press. pp 197-203
42. Goldstein AH (1995) Recent progress in understanding the molecular genetics and biochemistry of calcium phosphate solubilization by gram negative bacteria. Biol Agric Hortic 12:185-193
43. Goldstein AH, Liu ST (1987) Molecular cloning and regulation of a mineral phosphate solubilizing gene from *Erwinia herbicola*. Biotechnology 5:72-74

44. Grierson PF, Comerford NB, Jokela EJ (1998) Phosphorus mineralization kinetics and response of microbial phosphorus to drying and rewetting in a Florida Spodosol. Soil Biol Biochem 30:1323-1331

45. Gunes A, Ataoglu N, Turan M, Esitken A, Ketterings QM (2009) Effects of phosphate-solubilizing microorganisms on strawberry yield and nutrient concentrations. J Plant Nutr Soil Sci 172:385-392

46. Gupta RR, Singal R, Shanker A, Kuhad RC, Saxena RK (1994) A modified plate assay for secreening phosphate solubilizing microorganisms. J Gen Appl Microbiol 40:255-260

47. Gyaneshwar P, Naresh KG, Parekh LJ, Poole PS (2002) Role of soil microorganisms in improving P nutrition of plants. Plant Soil 245:83-93

48. Hamdali H, Bouizgarne B, Hafidi M, Lebrihi A, Virolle MJ, Ouhdouch Y (2008) Screening for rock phosphate solubilizing Actinomycetes from Moroccan phosphate mines. Appl Soil Ecol 38:12-19

49. Hamdali H, Hafidi M, Virolle MJ, Ouhdouch Y (2008) Growth promotion and protection against damping-off of wheat by two rock phosphate solubilizing actinomycetes in a P-deficient soil under greenhouse conditions. Appl Soil Ecol 40:510-517

50. Havlin J, Beaton J, Tisdale SL, Nelson W (1999) Soil fertility and fertilizers. An introduction to nutrient management. Upper Saddle River, NJ: Prentice Hall.

51. He ZL, Wu J, O'Donnell AG, Syers JK (1997) Seasonal responses in microbial biomass carbon, phosphorus and sulphur in soils under pasture. Biol Fertil Soils 24:421-428 t

52. Hwangbo H, Park RD, Kim YW, Rim YS, Park KH, Kim TH, Such JS, Kim KY (2003) 2-ketogluconic acid production and phosphate solubilization by Enterobacterintermedium. Curr Microbiol 47:87-92

53. Illmer PA, Schinner F (1992) Solubilization of inorganic phosphates by microorganisms isolated from forest soil. Soil Biol Biochem 24:389-395 Publisher Full Text

54. Illmer PA, Schinner F (1995) Solubilization of inorganic calcium phosphates solubilization mechanisms. Soil Biol Biochem 27:257-263

55. Illmer PA, Barbato A, Schinner F (1995) Solubilization of hardly soluble AlPO4 with P-solubilizing microorganisms. Soil Biol Biochem 27:260-270

56. Isherwood KF (2000) Mineral Fertilizer Use and the Environment. Paris: International Fertilizer Industry Association/United Nations Environment Programme.

57. Jacobs H, Boswell GP, Ritz K, Davidson FA, Gadd GM (2002) Solubilization of calcium phosphate as a consequence of carbon translocation by *Rhizoctonia solani*. FEMS Microbiol Ecol 40:65-71

58. Jorquera MA, Hernandez MT, Rengel Z, Marschner P, Mora MD (2008) Isolation of culturable phosphor bacteria with both phytate-mineralization and phosphate-solubilization activity from the rhizosphere of plants grown in a volcanic soil. Biol Fertil Soils 44:1025-1034

59. Jorquera MA, Crowley DE, Marschner P, Greiner R, Ferna'ndez MT, Romero D, Menezes-Blackburn D, De La Luz Mora M (2011) Identification of b-propeller phytase-encoding genes in culturable Paenibacillus and bacillus sp. from the rhizosphere of pasture plants on volcanic soils. FEMS Microbiol Ecol 75:163-172

60. Juma NG, Tabatabai MA (1977) Effects of trace-elements on phosphatase-activity in soils. Soil Sci Soc Am J 41:343-346

61. Juma NG, Tabatabai MA (1998) Hydrolysis of organic phosphates by corn and soybean roots. Plant Soil 107:31-38

62. Khan MR, Khan SM (2002) Effect of root-dip treatment with certain phosphate solubilizing microorganisms. Bioresour Technol 85(2):213-215

63. Khan AA, Jilani G, Akhtar MS, Naqvi SMS, Rasheed M (2009) Phosphorus solubilizing bacteria: occurrence, mechanisms and their role in crop production. J Agric Biol Sci 1(1):48-58

64. Khan MS, Zaidi A, Wani PA, et al. (2009) Role of phosphate solubilising microorganisms in sustainable agriculture. In: Lictfouse E (ed) Sustainable Agriculture, springer. p 552 DOI: 10.1007/978-90-481-2666-8_34

65. Khan MS, Zaidi A, Wani PA, Oves M (2009) Role of plant growth promoting rhizobacteria in the remediation of metal contaminated soils. Environ chem lett 7:1-19

66. Khan MS, Zaidi A, Ahemad M, Oves M, Wani PA (2010) Plant growth promotion by phosphate solubilizing fungi – current perspective. Arch Agron Soil Sci 56:73-98
67. Kim KY, McDonald GA, Jordan D (1997) Solubilization of hydroxyapatite by *Enterobacter agglomerans* and cloned Escherichia coli in culture medium. Biol Fertil Soils 24:347-352
68. Kim KY, Jordan D, McDonald GA (1998) *Enterobacter agglomerans*, phosphate solubilizing bacteria, and microbial activity in soil: effect of carbon sources. Soil Biol Biochem 30:995-1003
69. Krishnaraj PU, Khanuja SPS, Sadashivam KV (1998) Mineral phosphate solubilization (MPS) and mps genes -components in eco-friendly P fertilization. Bangalore: Abstracts of Indo US Workshop on Application of Biotechnology for Clean Environment and Energy, National Institute of Advanced Studies. p 27
70. Kucey RMN (1983) Phosphate solubilizing bacteria and fungi in various cultivated and virgin Alberta soils. Can J Soil Sci 63:671-678
71. Kumar V, Behl RK, Narula N (2001) Establishment of phosphate-solubilizing strains of *Azotobacter chroococcum* in the rhizosphere and their effect on wheat cultivars under greenhouse conditions. Microbiol Res 156:87-93
72. Lindsay WL, Vlek PLG, Chien SH (1989) Phosphate minerals. In: Dixon JB, Weed SB (eds) Minerals in soil environment, 2nd edn. USA. pp: Soil Science Society of America, Madison, WI. pp 1089-1130
73. Lopez BR, Bashan Y, Bacilio M (2011) Endophytic bacteria of *Mammillaria fraileana*, an endemic rock-colonizing cactus of the Southern Sonoran Desert. Arch Microbiol 193:527-541
74. Maliha R, Samina K, Najma A, Sadia A, Farooq L (2004) Organic acids production and phosphate solubilization by phosphate solubilizing microorganisms under in vitro conditions. Pak J Biol Sci 7:187-196
75. McGill WB, Cole CV (1981) Comparative aspects of cycling of organic C, N, S and P through soil organic matter. Geoderma 26:267-268
76. Nannipieri P, Giagnoni L, Landi L, Renella G (2011) Role of phosphatase enzymes in soil. In: Bunemann E, Oberson A,

Frossard E (eds) Phosphorus in action: Biological processes in soil phosphorus cycling. Soil biology, 26, Heidelberg: Springer. pp 251-244

77. Nautiyal CS (1999) An efficient microbiological growth medium for screening of phosphate solubilizing microorganisms. FEMS Microbiol Lett 170:265-270

78. Norrish K, Rosser H (1983) Mineral phosphate. In: Soils: an Australian viewpoint. Melbourne, CSIRO/London, UK, Australia: Academic Press. pp 335-361

79. Oliveira CA, Sa NMH, Gomes EA, Marriel IE, Scotti MR, Guimaraes CT, Schaffert RE, Alves VMC (2009) Assessment of the mycorrhizal community in the rhizosphere of maize (Zea mays L.) genotypes contrasting for phosphorus efficiency in the acid savannas of Brazil using denaturing gradient gel electrophoresis (DGGE). Appl Soil Ecol 41:249-258

80. Omar SA (1998) The role of rock phosphate solubilizing fungi and vesicular arbuscular mycorrhiza (VAM) in growth of wheat plants fertilized with rock phosphate. World J Microbiol Biotechnol 14:211-219

81. Park KH, Lee CY, Son HJ (2009) Mechanism of insoluble phosphate solubilization by Pseudomonas fluorescens RAF15 isolated from ginseng rhizosphere and its plant growth-promoting activities. Lett Appl Microbiol 49:222-228

82. Parker DR, Reichmann SM, Crowley DE (2005) Metal chelation in the rhizosphere. In: Zobel RW (ed) Roots and soil management: interactions between roots and the soil. Agronomy monograph no. 48, Madison: American Society of Agronomy. pp 57-93

83. Parks EJ, Olson GJ, Brinckman FE, Baldi F (1990) Characterization by high performance liquid chromatography (HPLC) of the solubilization of phosphorus in iron ore by a fungus. J Ind Microbiol Biotechnol 5:183-189

84. Patil MG, Sayyed RZ, Chaudhari AB, Chincholkar SB (2002) Phosphate Solubilizing Microbes: A Potential Bioinoculant for Efficient Use of Phosphate Fertilizers. In: Reddy SM, Reddy SR, Grisham S (eds) Bioinoculants for Sustainable Agriculture and Forestry, Jodhpur: Scientific Publisher. pp 107-118

85. Peix A, Velazquez E, Martýnez-Molina E (2007) Molecular methods for biodiversity analysis of phosphate solubilizing microorganisms (PSM). In: Velazquez E, Rodrguez-Barrueco C (eds) First international meeting on microbial phosphate solubilization, Berlin: Springer. pp 97-100
86. Pikovskaya RI (1948) Mobilization of phosphorus in soil in connection with vital activity of some microbial species. Microbiology 17:362-370
87. Puente ME, Bashan Y, Li CY, Lebsky VK (2004) Microbial populations and activities in the rhizoplane of rock-weathering desert plants. I. Root colonization and weathering of igneous rocks. Plant Biol 6:629-642
88. Puente ME, Li CY, Bashan Y (2004) Microbial populations and activities in the rhizoplane of rock-weathering desert plants. II. Growth promotion of cactus seedlings. Plant Biol 6:643-650
89. Puente ME, Li CY, Bashan Y (2009) Rock-degrading endophytic bacteria in cacti. Environ Exp Bot 66:389-401
90. Puente ME, Li CY, Bashan Y (2009) Endophytic bacteria in cacti seeds can improve the development of cactus seedlings. Environ Exp Bot 66:402-408
91. Quiquampoix H, Mousain D (2005) Enzymatic hydrolysis of organic phosphorus. In: Turner BL, Frossardand E, Baldwin DS (eds) Organic phosphorus in the environment, Wallingford UK: CAB International. pp 89-112
92. Renella G, Egamberdiyeva D, Landi L, Mench M, Nannipieri P (2006) Microbial activity and hydrolase activities during decomposition of root exudates released by an artificial root surface in Cd-contaminated soils. Soil Biol Biochem 38:702-708
93. Rengel Z, Marschner P (2005) Nutrient availability and management in the rhizosphere: exploiting genotypic differences. New Phytology 168:305-312
94. Reyes I, Bernier L, Simard RR, Antoun H (1999) Effect of nitrogen source on the solubilization of different inorganic phosphates by an isolate of *Penicillium rugulosum* and two UV-induced mutants. FEMS Microbiol Ecol 28:281-290
95. Reyes I, Baziramakenga R, Bernier L, Antoun H (2001) Solubilization of phosphate rocks and minerals by a wild type

strain and two UV induced mutants of *Penicillium regulosum*. Soil Biol Biochem 33:1741-1747

96. Reyes I, Bernier L, Antoun H (2002) Rock phosphate solubilization and colonization of maize rizosphere by wild and genetically modified strains of *Penicillium rugulosum*. Microb Ecol 44:39-48
97. Richardson AE (1994) Soil microorganisms and phosphorus availability. In: Pankhurst CE, Doubeand BM, Gupta VVSR (eds) Soil biota: management in sustainable farming systems, Victoria, Australia: CSIRO. pp 50-62
98. Richardson AE (2001) Prospects for using soil microorganisms to improve the acquisition of phosphorus by plants. Aust J Plant Physiol 28:897-906
99. Richardson AE, Simpson RJ (2011) Soil microorganisms mediating phosphorus availability. Plant Physiol 156:989-996
100. Richardson AE, Hadobas PA, Hayes JE, O'Hara CP, Simpson RJ (2001) Utilization of phosphorus by pasture plants supplied with myo-inositol hexaphosphate is enhanced by the presence of soil microorganisms. Plant Soil 229:47-56
101. Richardson AE, Barea JM, McNeill AM, Prigent-Combaret C (2009) Acquisition of phosphorus and nitrogen in the rhizosphere and plant growth promotion by microorganisms. Plant Soil 321:305-339Publisher Full Text
102. Richardson AE, Hocking PJ, Simpson RJ, George TS (2009) Plant mechanisms to optimize access to soil phosphorus. Crop Pasture Sci 60:124-143
103. Rodriguez H, Fraga R (1999) Phosphate solubilizing bacteria and their role in plant growth promotion. Biotechnol Adv 17:319-339
104. Rodriguez H, Fraga R, Gonzalez T, Bashan Y (2006) Genetics of phosphate solubilization and its potential applications for improving plant growth-promoting bacteria. Plant Soil 287:15-21
105. Rudolph W (1922) Influence of S oxidation upon growth of soybeans and its effect on bacterial flora of soil. Soil Sci l4:247-263
106. Saber K, Nahla LD, Chedly A (2005) Effect of P on nodule formation and N fixation in bean. Agron Sustain Dev 25:389-393

107. Sanchez P, Logan T (1992) Myths and science about the chemistry and fertility of soils in the tropics. In: Lal R, Sanchez P (eds) Myths and science of soils of the tropics, Madison, WI: Soil Science Society of America. pp 35-46

108. Schindler DW, Hecky RE, Findlay DL, Stainton MP, Parker BR, Paterson MJ, Beaty KG, Lyng M, Kasian SEM (2008) Eutrophication of lakes cannot be controlled by reducing nitrogen input: results of a 37-year whole-ecosystem experiment. Proc Natl Acad Sci U S A 105:11254-11258

109. Sims JT, Pierzynski GM (2005) Chemistry of phosphorus in soil. In: Tabatabai AM, Sparks DL (eds) Chemical processes in soil, SSSA book series 8, Madison: SSSA. pp 151-192

110. Singal R, Gupta R, Saxena RK (1994) Rock phosphate solubilization under alkaline conditions by*Aspergillus japonicus* and *A. foetidus*. Folia Microbiol 39:33-36

111. Song OR, Lee SJ, Lee YS, Lee SC, Kim KK, Choi YL (2008) Solubilization of insoluble inorganic phosphate by *Burkholderia cepacia* Da23 isolated from cultivated soil. Braz J Microbiol 39:151-156

112. Sperber JI (1958) The incidence of apatite-solubilizing organisms in the rhizosphere and soil. Aust J Agr Res 9:778-781

113. Sperber JI (1958) Solubilization of apatite by soil microorganisms producing organic acids. Aust J Agr Res 9:782-787

114. Sridevi M, Mallaiah KV, Yadav NCS (2007) Phosphate solubilization by *Rhizobium* isolates from Crotalaria species. J Plant Sci 2:635-639

115. SubbaRao NS (1982) Advances in agricultural microbiology. India: Oxford and IBH Publications Company. pp 229-305

116. Sutherland IW (2001) Biofilm exopolysaccharides: a strong and sticky framework. Microbiology 147:3-9

117. Swaby R, Sperber JI (1958) Phosphate Dissolving Microorganisms in the Rhizosphere of Legume, Nutrition Of Legumes; Proc. Univ. Nottingham 5^{Th} Easter Sch. Agril. Sci. (CSIRO Adelaide). Soils & Fert 22, 286(1959):289-294

118. Taha SM, Mahmoud SAZ, El-Damaty AA, Abd El- Hafez AM (1969) Activity of phosphate dissolving bacteria in Egyptian soil. Plant Soil 31:149

119. Tarafdar JC, Yadav RS, Meena SC (2001) Comparative efficiency of acid phosphatase originated from plant and fungal sources. J Plant Nutr Soil Sci 164:279-282

120. Tilman D, Fargione J, Wolff B, D'Antonio C, Dobson A, Howarth R, Schindler D, Schlesinger WH, Simberloff D, Wackhamer D (2001) Forecasting agriculturally driven global environmental change. Science 292:281-284

121. Torsvik V, Ovreas L (2002) Microbial diversity and function in soil: from genes to ecosystems. Curr Opin Microbiol 5:240-245

122. Trolove SN, Hedley MJ, Kirk GJD, Bolan NS, Loganathan P (2003) Progress in selected areas of rhizosphere research on P acquisition. Aust J Soil Res 41:471-499

123. Vassilev N, Vassileva M, Azcon R, Medina A (2001) Preparation of gel-entrapped mycorrhizal inoculum in the presence or absence of *Yarrowia lypolytica*. Biotechnol Lett 23:907-909

124. Vassilev N, Vassileva M, Nikolaeva I (2006) Simultaneous P-solubilizing and biocontrol activity of microorganisms: potentials and future trends. Appl Microbiol Biotechnol 71:137-144

125. Vazquez P, Holguin G, Puente M, Lopez-cortes A, Bashan Y (2000) Phosphate solubilizing microorganisms associated with the rhizosphere of mangroves in a semi-arid coastal lagoon. Biol Fertil Soils 30:460-468

126. Venkateswarlu B, Rao AV, Raina P, Ahmad N (1984) Evaluation of phosphorus solubilization by microorganisms isolated from arid soil. J Indian Soc Soil Sci 32:273-277

127. Vyas P, Gulati A (2009) Organic acid production in vitro and plant growth promotion in maize under controlled environment by phosphate-solubilizing fluorescent *Pseudomonas*. BMC Microbiol 9:174

128. Wakelin SA, Warren RA, Harvey PR, Ryder MH (2004) Phosphate solubilization by *Penicillium* sp. closely associated with wheat roots. Biol Fertil Soils 40:36-43

129. Wani PA, Zaidi A, Khan AA, Khan MS (2005) Effect of phorate on phosphate solubilization and indole acetic acid (IAA) releasing potentials of rhizospheric microorganisms. Annals Plant Protection Sci 13:139-144

130. Whitelaw MA (2000) Growth promotion of plants inoculated with phosphate solubilizing fungi. Adv Agron 69:99-151

131. Whitelaw MA, Harden TJ, Helyar KR (1999) Phosphate solubilization in solution culture by the soil fungus *Penicillium radicum*. Soil Biol Biochem 32:655-665

132. Widada J, Damarjaya DI, Kabirun S (2007) The interactive effects of arbuscular mycorrhizal fungi and rhizobacteria on the growth and nutrients uptake of sorghum in acid soil. In: Rodriguez-Barrueco C (ed) Velazquez E, Springer: First International Meeting on Microbial Phosphate Solubilization. pp 173-177

133. Yi Y, Huang W, Ge Y (2008) Exopolysaccharide: a novel important factor in the microbial dissolution of tricalcium phosphate. World J Microbiol Biotechnol 24:1059-1065

134. Zaidi A, Khan MS, Ahemad M, Oves M, Wani PA, et al. (2009) Recent Advances in Plant Growth Promotion by Phosphate-Solubilizing Microbes. In: Khan MS (ed) Microbial Strategies for Crop Improvement, Berlin Heidelberg: Springer-Verlag. pp 23-50

135. Zhou K, Binkley D, Doxtader KG (1992) A new method for estimating gross phosphorus mineralization and immobilization rates in soils. Plant Soil 147:243-250

136. Zhu F, Qu L, Hong X, Sun X (2011) Isolation and characterization of a phosphate-solubilizing halophilic bacterium *Kushneria* sp. YCWA18 from Daqiao Saltern on the coast of Yellow Sea of China. Evid base Compl Alternative Med 615032:6

Chapter 8

Use of Vermicomposting Biotechnology for Recycling Organic Wastes in Agriculture

Gunindra Nath Chattopadhyay

Institute of Agriculture, Visva-Bharati University, Sriniketan, West Bengal 731235, India

ABSTRACT

Recent trend of declining sustainability in agricultural production is appearing as a major threat to most of the Asian countries. To combat the situation, increasing importance is now being paid on incorporation of organic materials for rejuvenating the health condition and, hence, the productivity of these soils. Large scale availability of conventional organic manures being a big problem nowadays, major attention is being paid on recycling of different kinds of organic wastes for this purpose. Apart from abatement of environmental pollution, such reuse of organic wastes in agriculture helps in the improvement of various physical, chemical and biological properties of the soils and, thus, helps in sustaining the soil health. While recycling different organic wastes in agriculture, it is essential to process the materials through an adequate period of composting for improving their usability in the soils. However, due to some short comings of traditional composting systems, the technology of recycling of organic wastes has not been widely accepted so far. Under this situation, vermicomposting has recently emerged as a simple but efficient biotechnology for recycling wide ranges of organic wastes with the help of some specific groups of earthworms. In view of the growing popularity of this biotechnology, various aspects of waste recycling in agriculture through vermicomposting have been dealt in this communication.

REVIEW

Introduction

With the rapidly increasing population in most of the Asian countries, the major challenge facing the mankind during the beginning of the new millennium is to provide food sustenance for all the people of this continent. With per capita land area decreasing consistently, various measures are being adopted to increase the agricultural production from these shrinking resources to meet the emergent demand of escalating population. However, sustaining the food production from these decreasing land areas depends largely on one factor, maintenance of soil health at high levels for encouraging good growth of plants. The recent trend of consistently reducing use efficiency of mineral fertilizers under high productive systems associated with the problems of gradual deterioration of soil health due to indiscriminate use of fertilizers is raising frequent questions about over dependence on mineral fertilizers in sustaining the health and, consequently, the productivity of the arable soils of this continent. It has been appreciated that mineral fertilizers can only supply plant nutrients to the soils but they cannot take care of other physical, chemical and biological attributes of soil health. On the other hand, organic materials play a much more positive role in this respect. Under this context, the concept of integrated plant nutrition system involving both inorganic and organic sources of nutrients has been conceived and gaining momentum ([Chattopadhyay 2005]). Such integration of mineral fertilizers and organic manures has special significance for many of the Asian countries because the arable soils of most tropical and subtropical countries are poor in organic matter due to high temperature and more intense microbial activity ([Gaur 2006]). Hence, a regular and sizeable addition of organic material to soil is essential for maintaining optimum organic matter status and, thereby, sustaining the health of the soil ([Manna et al. 2005]).

The beneficial roles of organic matter in the sustenance of soil health and productivity are well documented ([Kumazawa 1984]). [Allison] ([1973]) summarized the important contributions of organic matter in sustaining the productivity levels of soils to be as follows:

1. It is a natural resource of plant nutrients and microbial energy.
2. It serves as cation exchange site and chelating agent.
3. It improves physical condition of soils.

However, in spite of such significant importance of using organic matter in tropical soils, large-scale availability of organic materials from traditional sources has remained a big problem. It is, therefore, necessary to utilize effectively all the organic resources available on and off the farm. Under this context, the potentials of reusing various kinds of organic wastes have been emphasized by many workers ([Gallardo-Lara] and [Nogales][1987]; [Howe] and [Coker][1992]).[Gupta] et al. ([1998]) calculated that, on an average, around 4 to 5 kg of organic wastes are generated per person per day in India. This calculation is likely to be by and large applicable to most of the Asian countries also. Thus, with escalation of human population, huge amount of varying natures of organic wastes are being generated in these countries every year. [Bisoyi] ([2003]) made a systematic assessment of the quality of some common organic wastes in India and has shown these wastes to be the source of large amount of nutrient elements in addition to organic matter. Successful recycling of these wastes in different Asian countries will not only add substantial amount of all the plant nutrients but will also bring in the much-needed organic matter to the soils, leading to betterment of soil health (Table 1).

Table 1: Potentials of NPK from different organic wastes in India

Source	Plant nutrients (million ton)			
	N	P 2 O 5	K 2 O	Total
Cattle	2.997	0.793	1.332	5.102
Buffalo	0.745	0.276	0.487	1.508
Goat and sheep	0.214	0.063	0.020	0.297
Pig	0.044	0.027	0.029	0.100
Poultry	0.027	0.020	0.010	0.057
Other livestock	0.079	0.018	0.069	0.166
Human beings	3.228	0.776	0.715	4.719
Farm crop wastes	5.600	2.300	10.700	18.600
Forest litter	0.075	0.030	0.075	0.180
Water hyacinth compost	0.060	0.033	0.075	0.168
Rural compost	1.130	0.678	1.130	2.938
Urban compost	0.024	0.015	0.030	0.069
Sewage sludge	0.012	0.009	0.003	0.024
Total	*14.215*	*5.038*	*14.675*	*39.928*

Adapted from [Bisoyi (2003]).
Chattopadhyay

While recycling different organic wastes in agriculture, it is essential to process them through an adequate period of composting. Composting has been described as a biological process for converting solid waste into a stable, humus-like product, which is used as a soil conditioner ([Talashilkar][1989]). Apart from providing substantial plant nutrition, these composts promote soil aggregation, improve air-water relationship, enhance cation exchange capacity, increase water retentivity, provide energy and improve several other physico-chemical properties of soils ([Webber][1978]; [Epstein][1997]). In addition, importance of composts in taking care of most of the pathogens present in the substrates has been emphasized by [Tandon] (.[1995]). [Chen] and[Wu] ([2005]) discussed wide ranges of benefits from use of composts in different agricultural systems. However, the major constraints related to large scale adoption of composting technology have been described to be their bulk volume, long time required for composting, incomplete decomposition of resistant components, lower nutrient values as compared to mineral fertilizers and, above all, the complicated processes of composting ([Chattopadhyay 2005]). Good amount of studies have been and are being carried out at different levels to overcome these problems. In the course of these studies, in recent years, vermicomposting has emerged as an effective biotechnology for decomposing wide ranges of organic wastes with the help of intestinal microorganisms of some specific kinds of earthworms ([Edwards] and [Lofty][1972]; [Kale 1993]). Composts produced with such biotechnology have been found to be superior in nutrient status than the traditionally prepared compost and to contain various vitamins, plant growth regulators, antibiotics, etc., ([Tilak] et al. [2010]) resulting in better performance after their field application. In the present communication, the possibility of recycling different organic wastes in agriculture through vermicomposting biotechnology has been discussed.

Vermicomposting Biotechnology

Beneficial effects of earthworms in improving structure, aeration, nutrient status and some other properties of the soils and, thereby, the growth of the crops have been known since long ([Darwin][1837]). However, the knowledge about the efficiency of some groups of the earthworms in decomposing various organic materials was gathered later on, and the concept of utilizing this behavior for composting wide ranges of organic wastes was conceived during mid-twentieth century ([Senapati 1993]). Based on ecological niches, earthworms may be broadly grouped into three categories *viz.* endogeics, anecics and epigeics ([Bouche 1977]). Endogeic earthworms are geophagous in nature, which live deep inside the soil. Anecics reside just below the soil surface and feed on the organic materials mixed with the soils. The third group is known as epigeics which live on upper surface of soils feeding mainly on plant litter and other organic debris available on the soil surface. As these earthworms can consume a variety of organic matters, they are most suitable for converting organic wastes into useful organic manures. However, uses of anecic and endogeic earthworms in such composting process have also been reported by [Lavelle] and [Martin] ([1992]). A list of common earthworm species suitable for vermicomposting has been presented in Table 2.

These epigeic earthworms being voracious feeders can consume large quantities of different organic materials. In spite of their high rate of consumption, they utilize only a very small part of the consumed food for their body synthesis and excrete about 90% to 95% of the ingested materials as vermicast. Since the intestines of the earthworms harbor high concentrations of various kinds of microorganisms ([Wallwork 1984]), these vermicasts exhibit rich population of these organisms also ([Atlavinyte] and [Vanagas][1982]). Analysis of gut content of the earthworms has revealed the occurrence of different kinds of symbionts like microfungi, bacteria, protozoans, actinomycetes, etc. ([Dash] and [Senapati][1986]). Earthworm guts provide congenial environmental condition for their proliferation, and hence, the total numbers of such microorganisms in earthworm intestines generally increase many folds as compared to their habitats ([Ponomarera][1962]). The food materials ingested by the earthworms are, thus, subjected to more intense microbial activity in the alimentary canal not only due to these higher concentrations of different microorganisms but also owing to increased surface area of the food particles meshed to very fine sizes while passing through the gizzard. In addition to the activities of the microbes, occurrence of various enzymes in earthworm guts also helps such processes. These enzymes operate in near neutral to slightly alkaline pH range in the intestinal fluid of the worms ([Senapati 1993]). In addition, existence of various hormones, antibiotics, etc. has also been reported in earthworm gut ([Edwards] and [Lofty][1972]). All these components mix thoroughly with the food materials in the intestines of the earthworms and are released with their excreta to be known as vermicast. These vermicasts undergo rapid aerobic decomposition in presence of different microbes, which activity is accelerated by various enzymes and encouraged by occurrence of numbers of growth promoting substances. Aerobic microorganisms, being more active decomposers than the anaerobes, degrade vermicasts more rapidly than the traditional 'pit composting' methods, and a nutrient rich well-humified organic manure is obtained in shorter period of time which is termed as 'vermicompost' ([Chattopadhyay 2005]).

Table 2: List of some earthworm species suitable for vermicomposting

Lumbricidae	**Megascolecidae**	**Ocnerodrilidae**	**Octaochaetidae**	**Eudrilidae**	**Moniligastridae**
Bimasto parvus	*Lampito mauritii*	*Ocnerodrulus occidenalis*	*Dichogaster bolaui*	*Eudrilus euginae*	*Moniligaster perrieri*
Eisenia foetida	*Perionyx excavatus*		*Dichogaster saliens*		
Eisenia hortensis	*Metaphire anomala*		*Romiella bishambari*		
	Polypheretima elongate				

Screened from [Julka 1988].

Chattopadhyay

Benefits of Vermicomposting

As discussed earlier, vermicomposting biotechnology helps in recycling different organic wastes with the help of large numbers of aerobic microorganisms. This results, in general, better nutrient status of the produced materials, as compared to those prepared by traditional composting systems. [Scot] ([1998]), while assessing the nature and properties of cattle wastes, digested with and without earthworms, observed availability of nitrogen to be higher in vermicomposted materials. While working on nature and magnitudes of transformation of phosphorus during the course of vermicomposting of different wastes, [Ghosh] et al. ([1999b]) observed that vermicomposting process tends to reduce the quantum of fixation of P as insoluble Fe, Al and Ca phosphate, and also helped mineralize P from organic to inorganic forms. Among the effects of different microorganisms and enzymes contributing to such increased availability of phosphorus, major emphasis may be given to the presence of very high concentration of phosphate-solubilizing bacteria in the vermicast. [Bhattacharya] and [Chattopadhyay] ([2002]) reported vermicasts prepared from mixtures of cattle wastes and fly ash to exhibit high occurrence of such microorganism. Such rich occurrence of phosphorus-solubilizing bacteria is likely to solubilize good amount of phosphorus from insoluble form in any organic waste material, and also to qualify vermicompost as a potential phosphatic biofertilizer. Increased nutrient availability and microbial activity due to vermicomposting, as reported by [Jambhekar] ([1992]), has been presented in Table 3. [Ghosh] et al. ([1999a]) also reported the results of a study on effect of inclusion of epigeic earthworm *Eisenia foetida* on several organic wastes *viz.* cow dung, poultry droppings, kitchen wastes, municipal wastes and dry leaves, and observed that the availability of three major nutrient elements *viz.* N, P and K, and four micronutrients *viz.* Fe, Cu, Mn and Zn was considerably higher in the vermicomposted treatments for all the wastes as compared to those without earthworms. Similarly, adoption of vermicomposting technology has shown to increase the concentration of beneficial bacteria like nitrogen-fixing ones too ([Bhattacharya] and[Chattopadhyay][2004]). This helps in accumulation of atmospheric N_2 into the vermicomposted waste materials, thus, increasing their nitrogen status. The behavior is likely to facilitate the composting of different slowly degradable waste materials like saw dust, coir waste, sugarcane trash, etc. which exhibit high lignocellulose content, wide C:N ratio and low moisture.

Table 3: Status of major nutrients and microbial population in some organic wastes treated with and without earthworms

Substrate	**Parameters**				
	C:N ratio	**N (%)**	**Available P 2 O 5 (%)**	**Available K 2 O (%)**	**Microbial count**
Cow dung					
Original	49.1	0.53	0.003	0.104	-
Vermicomposted	16.8	1.20	0.004	0.396	24×106
Untreated	24.0	1.00	0.003	0.192	18×105
Sugarcane trash					
Original	38.17	0.66	0.080	0.020	-
Vermicomposted	16.07	1.40	0.100	0.010	24×106
Untreated	30.72	0.80	0.090	0.060	20×105
Horse manure					
Original	55.3	0.68	0.010	0.060	-
Vermicomposted	15.3	1.80	0.030	0.020	30×106
Untreated	29.6	1.29	0.020	0.010	28×105

Adapted from [Jambhekar (1992]).

Chattopadhyay

Such high nutrient content and other beneficial properties of vermicompost help in increased production of healthier crops. As shown in Table 4, substantial yield increments in different crops on integration of chemical fertilizers with vermicompost prepared from organic wastes have been reported by [Ghosh] et al. ([1999a]). Similar results on different vegetables have also been observed under an ongoing research program at this center. All these studies have shown use of vermicompost to be able to reduce the application of mineral fertilizers substantially without affecting the crop yield.

Table 4: Effect of vermicompost on crop yields (t/ha)

Treatments	Potato	Mustard	Mulberry	Marigold
FYM+NPK100	36.8	1.6	8.1	9.6
VC+NPK100	44.3	1.9	10.5	11.6
VC+NPK75	43.5	1.87	9.7	13.4
VC+NPK50	43.0	1.89	9.7	16.3
Control	12.3	0.40	2.7	4.2

FYM, farm yard manure; VC, vermicompost; $NPK_{100,75,50}$, percent of recommended NPK fertilization (Ghosh et al. [1999a]).

Chattopadhyay

Occurrences of different diseases in the crops have also been found to come down due to use of vermicompost, thus, reducing the need of pesticide application to the plants. While working on use of vermicomposted fly ash in agriculture, [Iftikar] et al. ([2011]) reported incidence of different diseases to be reduced substantially on use of vermicompost in tomato cultivation (Table 5). Similar reports are available for other crops also.

Table 5: Disease infestation of tomato under different treatments

Protocol for vermicomposting

Treatments	Bacterial wilt	Leaf curl	Bacterial blight
FYM 10 tha-1+NPK 100%	2	4	6
VC 10 tha-1+NPK 100%	1	1	2
VF 10 tha-1+NPK 100%	1	2	2
NPK 100%	6	11	14

FYM, farm yard manure; VC, vermicompost; VF, vermicomposted fly ash; NPK_{100}, percent of recommended NPK fertilization.

Chattopadhyay

Protocol for Vermicomposting

By following the conventional system of composting, vermicomposting is generally undertaken in small pits. However, recent studies carried out at Institute of Agriculture, Visva-Bharati University have shown vermicomposting should preferably be carried out above ground level, in a place without possibility of water stagnation ([Chattopadhyay 2005]). This will ensure occurrence of aerobic condition. The bed may be enclosed with brick-made structures with good numbers of holes or the wastes be kept as heaps without any enclosure. A shed is to be provided above the bed to prevent entry of rainwater or sunshine.

Wide ranges of organic wastes may be used for vermicomposting. However, they should preferably be mixed with at least 20% to 25% cow dung or sprinkled with cow dung slurry. Large sized wastes like straws, plant twigs, etc. may be shredded to smaller sizes to facilitate composting. The mixture should be kept under moist condition for 10 to 15 days. This will allow the initial thermophilic condition of organic materials to subside to a tolerable level for the earthworms. This rise in temperature in the organic materials during primary period of decomposition helps kill most of the mesophilic pathogens in the waste materials, making the resultant compost more safe for field application.

After 10 to 15 days of primary decomposition, epigeic earthworms are to be incorporated in the organic materials at 8 to 10 worms kg^{-1} of waste. Selection of the species of earthworms for vermicomposting should primarily be based on their capacity for consumption of organic wastes, rate of multiplication and adaptability to local condition. Special care should be exercised to keep the wastes under moist condition through periodic addition of water during the entire course of composting. This will not only help the earthworms to get a favorable environment but will also permit more intensive activities of the microorganisms in degrading the wastes. Periodic bottom-up turning of the waste materials will be particularly useful for the composting process by exposing the lower layer of the waste materials to the aerobic zones. This will accelerate the rate of decomposition of the organic materials and also encourage multiplication of the earthworms. Temperature is also an important factor. Optimum temperature range for vermicomposting is around 25°C to 35°C. Rate of composting becomes slower under low temperature levels.

At maturation, the compost material assumes dark to brown coloration, becomes nonsticky and odorless. At this stage, the moisture content of the composted materials is brought down by stopping the application of water. The compost materials are then sieved, and the worms remaining above the sieve are transferred to another vermicompost bed which is kept ready with partially decomposed mixture of organic waste. Since the numbers of earthworms increase considerably during the period of vermicomposting, the amount of organic waste can be increased accordingly under successive compostings. In case of larger production of composts, sometimes high intensities of light are provided from the top of the compost heap. Since most earthworms do not like high intensity of light, they tend to migrate from upper surface to lower depth of the heap. Vermicompost may then be harvested from the top of the heap. However, sieving will ensure availability of uniform-sized granular particles. This will also reduce the possibility of less matured materials to be mixed with the compost.

CONCLUSIONS

In recent years, vermicomposting is emerging as a simple, easily adoptable and effective biotechnology for recycling wide ranges of organic wastes for agricultural production. This can be taken up in small scales at household levels or at large levels as business propositions. The former practice helps an agriculturist or gardener to produce vermicompost from his homestead garbage for the purpose of utilizing the generated vermicompost for own consumption. On the other hand, vermicomposting can also be taken up at commercial level using municipal and/or other sources of wastes and the product be marketed as good quality organic manure. The gradually expanding market of organic foods for sustaining human health presents significant opportunities for undertaking such activity on commercial basis. In this communication, a brief account of the concept, benefits and method of vermicomposting has been presented with relation to its agricultural use. Various processes of improvements of quality of vermicompost as well as pace of such composting are being suggested now by different workers. They include microbiological inoculation, enrichment of quality, improved management procedures, etc.([Bhattacharyya] and[Kumar][2009]; [Saha][2009]; [Mal][2010]). The vermicomposting technology may be modified suitably based on the demands of specific situation as well as requirement of the operator.

ACKNOWLEDGMENTS

Author thanks the members of IJROWA editing team for their patience with this manuscript.

REFERENCES

1. Allison FE (1973) Soil organic matter and its role in crop production. Elsevier Science Publication Co., New York.
2. Atlavinyte O, Vanagas J (1982) The effect of earthworms on the quality of barley and rye grain. Pedobiologia 23:255-362
3. Bhattacharyya P, Kumar D (2009) Enrichment of vermicompost with urea, rock phosphate, azolla, pressmud and biofertilisers. Agrobios Newsletter 8(2):39-40
4. Bhattacharya SS, Chattopadhyay GN (2002) Increasing bioavailability of phosphorus from fly ash through vermicomposting. J Environ Qual 31:2116-2119
5. Bhattacharya SS, Chattopadhyay GN (2004) Transformation of nitrogen during vermicomposting in fly ash. Waste Manag Res 22(6):488-491
6. Bisoyi RN (2003) Biofertiliser production technology: state-of-the art. In: Chand PK, Patro SN (eds) Science for society, Indian Science Congress Association, Bhubaneswar Chapter, Bhubaneswar. pp 87-92
7. Bouche MB (1977) Strategies lombriciennes. In: Lohm U, Persson T (eds) Soil organisms as components of ecosystems, Biol Bull, Stockholm. pp 122-132
8. Chattopadhyay GN (2005) Vermicomposting as a biotechnological tool for recycling organic wastes. In: Ghosh TK, Chakrabarti T, Tripathi G (eds) Biotechnology in environmental management, A.P.H. Publishing Corporation, New Delhi. pp 135-145

9. Chen J-H, Wu J-T (2005) Benefit and drawbacks of composting. In: Chen Z-S, Bejosano-Gloria C (eds) Compost production – a manual for Asian countries, FFTC, Taiwan. pp 2-7
10. Darwin C (1837) On the formation of vegetable mould, vol 2, series 5. Trans Geol Soc, London. pp 505-509
11. Dash MC, Senapati BK (1986) Vermitechnology, an option for organic waste management in India. In: Dash MC, Senapati BK, Mishra PC (eds) Verms and vermicomposting, Sambalpur University, Orissa, School of Life Sciences. pp 157-172
12. Edwards CA, Lofty JR (1972) Biology of earthworms. Chapman and Hall, London. p 283
13. Epstein E (1997) The Science of composting. Technomic Publising Co. Inc, USA. pp 383-415
14. Gallardo-Lara F, Nogales R (1987) Effect of the application of town refuse compost on the soil plant system- a review. Biological Wastes 19:35-62 Publisher Full Text
15. Gaur AC (2006) Biofertilizers in sustainable agriculture. Directorate of Information and Publications of Agriculture, ICAR, New Delhi. p 288
16. Ghosh M, Chattopadhyay GN, Baral K, Munsi PS (1999a) Possibility of using vermicompost for reconciling sustainability with productivity. In: Ghosh DC (ed) Proc. sem. agrotech. And environment, Visva-Bharati Univ, India. pp 64-68
17. Ghosh M, Chattopadhyay GN, Baral K (1999) Transformation of phosphorus through vermicomposting. Bio Resource Tech 69:149-154
18. Gupta SK, Sinvaran P, Sreenivasa R (1998) Management of rural and urban wastes in agriculture. In: Biswas TD, Narayanswami G (eds) Soil organic matter and organic residue management for sustainable productivity, Bull 19 Indian Soc Soil Sci, New Delhi. pp 135-153
19. Howe CA, Coker CS (1992) Co-composting municipal sewage sludge with leaves, yard wastes and other recyclables, a case study. In: Air Waste Management Association. 85th Annual Meeting and Exhibition. , Kansas City. 21–26 June 1992
20. Iftikar W, Roy G, Chattopadhyay GN (2011) Effect of vermicomposted fly ash on some agricultural crops. In: National

workshop on 'Use of vermicomposted fly ash in agriculture'. Visva-Bharati University, India. 26 May 2011

21. Jambhekar H (1992) Use of earthworms as a potential source of decomposed organic wastes. In: Ghosh TK, Chakrabarti T, Tripathi G (eds) Biotechnology in environmental management, A.P.H. Publishing Corporation, New Delhi. pp 135-145
22. Julka JM (1988) The fauna of India and adjacent countries. Zoological Survey of India, Calcutta. p 392
23. Kale RD (1993) Vermiculture - scope for new biotechnology. In: Ghosh AK (ed) Earthworms resources and vermiculture, Zoological Survey of India, Kolkata.
24. Kumazawa K (1984) Beneficial effects of organic matter on rice growth and yields in Japan. In: Organic matter and rice. IRRI, Philippines. pp 431-443
25. Lavelle P, Martin A (1992) Small scale and large scale effects of endogeic earthworms in soil organic matter dynamics in soils of the humid tropics. Soil Biol Biochem 24(12):1491-1498
26. Mal S (2010) Microbial management of vermicomposting technology. PhD Thesis. Visva-Bharati University, India. p 126
27. Manna MC, Swarup A, Wanjari RH, Ravankar HN, Mishra B, Saha MN, Singh YV, Sahi DK, Sarap PA (2005) Long-term effect of fertilizer and manure application on soil organic carbon storage, soil quality and yield sustainability under sub-humid and semi-arid tropical India. Field Crop Res 93:264-280
28. Ponomarera ST (1962) Soil macro and micro-organisms and their roles in increasing fertility. Vitoraya Zoologi Cheskaya Konferenciya Litovoskoi, SSR. pp 97-98
29. Saha A (2009) Effects of phosphorus enrichment on quality of vermicompost. M. Sc.(Ag.) Thesis. PSB, Visva-Bharati University, India.
30. Scot MA (1998) The use of worm digested animal waste as supplement to peat in loamless compost for hardy nursery stock. In: Edwards CE, Neuhauser EF (eds) Wastes and Environmental Management, SPB Acad. Publ, The Netherlands.
31. Senapati BA (1993) Earthworm gut contents and its significance. In: Ghosh AK (ed) Earthworm resources and vermiculture, Zoological Society of India, Kolkata. pp 97-100

32. Talashilkar SC (1989) Recycling of urban wastes in agriculture. In: Mishara PC (ed) Soil pollution and soil organism, Ashish Publ, House, New Delhi. pp 177-208
33. Tandon HLS (1995) Waste recycling in agriculture- an introduction. In: Tandon HLS (ed) Recycling of crop, animal, human and industrial wastes in agriculture, Fertilizer Dev. and Consultation Orgn, New Delhi. pp 1-8
34. Tilak KVBR, Pal KK, De R (2010) Microbes for sustainable agriculture. I.K. International Publ. House Pvt. Ltd., New Delhi, India. p 200
35. Wallwork JA (1984) Earthworm biology. 1st Indian edn. Arnold-Heinemann, New Delhi.
36. Webber LR (1978) Incorporation of non-segregated, non-composted solid wastes and soil physical properties. J Env Quality 7:397

Chapter 9

Liquid Microbial Consortium- A Potential Tool for Sustainable Soil Health

Pavan kumar Pindi[1] and SDV Satyanarayana[2]

[1]Associate professor, HOD, Department of Microbiology, Palamuru University, India [2]Department of Biotechnology, KL University, Guntur, A.P

ABSTRACT

The green revolution brought amazing consequences in food grain production but with insufficient concern for agricultural sustainability. The availability and affordability of fossil fuel based chemical fertilizers at farm level in India have been ensured only through imports and subsidies which are largely dependent on GDP of the country. Dependence on chemicals for future agricultural needs would result in further loss in soil health, possibilities of water contamination and calculated burden on the fiscal system. Indiscriminate synthetic fertilizer usage has polluted the soil, water basins, destroyed micro-organisms and eco-friendly insects, made the crop more susceptible to diseases and depleted soil fertility at the primary levels as of today, which is the main concern of the write up. In this critical context Microorganisms have been emerged as the potential alternative for the productivity, reliability and sustainability of the global food chain. Carrier based biofertilzers has already proved to be the best over the agro chemicals and have been showing the tremendous effect on the global agriculture productivity since the past two decades. Rectifying the disadvantages of the carrier based biofertilizers, liquid biofertilizers have been developed which would be the only alternative for the cost effective sustainable agriculture. The article focuses on Liquid Biofertilizer Technology providing reliable reasons for their necessity, specificity and emphasizes that "Use of agriculturally important microorganisms in different combinations i.e. Liquid microbial consortium (LMC) is the only solution for restoration of soil health".

Even though biofertilizers are being produced and distributed constantly by private agencies, NGO's, State and Central Government production units for the last three decades, their corresponding usage is not in the satisfactory proportions. To cope with the rising demands for food commodities, serious efforts are being made by the State and Central Governments (under the National Projects) for the sufficient agricultural production by popularizing biofertilizers and making them available to the farmer community. In spite of these efforts, the rate of consumption of biofertilizers is not to the optimum level in comparison with the agrochemicals. The reason attributed is the "non-availability of good and suitable carrier materials" that raises contamination problems and shorter shelf life. To cope with this alarming situation, Liquid formulations (LFs) are being developed that ensure more quality over the conventional carrier based biofertilizers inaugurating a new era in the Biological input technology. These liquid formulations facilitate long shelf life (up to 2 years), minimum contamination, carrier free activity, handling comfort, storage and transport convenience, easy quality control, enhanced export potentials and are preferred by the farmer community as well as manufacturers.

INTRODUCTION

Why to Explore Bio-Fertilizers?

India is an agricultural based country. In order to feed the ever growing populations, India has to increase the per unit area productivity. According to United Nations Food and Agriculture Organization (FAO) estimations, the average demand for the agricultural commodities will be 60% higher in 2030 than present time and more than 85% of this additional demand will be from developing countries [1]. For over half a century, the world has relied on the concept of increasing crop yields to supply an ever increasing demand of food. Therefore, vertical expansion of food production is necessary. In order to increase the unit area productivity of agricultural land, the role of different crop nutrients in contributing increased crop yield is vital. Among the crop nutrient, nitrogen as well as phosphorus play an important role in increasing the crop productivity. Further, the nitrogenous chemical fertilizers are manufactured industrially using nonrenewable petroleum products under high temperature and high pressure. Increase in petroleum cost day by day effects the cost of the chemical fertilizers. In addition, more than 50% of the applied N-fertilizers are somehow lost through different agricultural processes which not only lead to economical loss to the farmers and polluted environment consequently.

Feedstock/fossil fuels depletion and increasing fertilizer cost are making marginal farmer unaffordable. Growing concern about environmental hazards, increasing threat to sustainable agriculture are some of the other reliable reasons for the biofertilizer promotion. However, plant nutrients like N, P and K are highly essential for plant growth and Metabolism. It is also evident that plants utilizes nutrients in greater amounts from soil in modern intensive cultivation and needs replenishment. Under such conditions Microorganisms offer good alternative technology to replenish crop nutrients.

lubilizing/mobilizing/nutrient recycling. These microorganisms occur in soils naturally, but their populations are often scanty. In order to increase the crop yield, the desired microbes from rhizosphere are isolated and artificially cultured in adequate count and mixed with suitable carriers or as they are in suitable combinations (Microbial consortium) by artificial culturing. These are known as biofertilizers or microbial inoculants.

Estimated Demand and Supply of Biofertilizers

India is one of the important countries in biofertilizer production and consumption. In order to encourage the organic agriculture by biofertilizers, five biofertilizers namely Rhizobium, Azotobacter, Azospirillum, Phosphate solubilizing bacteria and mycorrhiza have been incorporated in the FCO, 1985 [2]. The average consumption in the country is about 45,000 tons per annum while the production being less than the half. At present in India there is a gap of about 10 million tones of plant nutrients between removal by crops and replenishment through fertilizers. It is evident that there is a tremendous gap between the annual demand and production of the biofertilizers globally especially in India. Hence, the judicious combination of chemical fertilizers and biofertilizers is also encouraged considering economical and ecological concerns. It is estimated that the present level of biofertilizer use is quite low and there is a substantial potential to increase it to 50,000-60,000 tons by 2020. The maximum production capacity is in Agro Industries Corporation followed by State Agriculture Departments, National Biofertilizers Development Centers, State Agricultural universities and private sectors.

LBF vs. CBBF

Liquid biofertilizer formulation is the promising and updated technology of the conventional carrier based production technology which inspite of many advantages over the agrochemicals, left a considerable dispute among the farmer community in terms of several reasons major being the viability of the organism. Shelf life is the first and foremost problem of the carrier based biofertilizers which is up to 3 months and it does not retain throughout the crop cycle, LBF on the other hand facilitates the long survival of the organism by providing the suitable medium which is sufficient for the entire crop cycle. Carrier based bio fertilizers are not so tolerant to the temperature which is mostly unpredictable and uncertain in the crop fields while temperature tolerance is the other advantage of the liquid biofertilizers. The range of possible contamination is very high as bulk sterilization does not provide the desirable results in the case of CBBF, where as the contamination can be controlled constructively by means of proper sterilization techniques and maintenance of intensive hygiene conditions by appropriate quality control measures in the case of LBF. Moisture retaining capacity of the CBBF is very low which does not allow the organism viable for longer period and the LBF facilitates the enhanced viability of the organism [3].

However, the administration of LBF in the fields is comparatively easier than CBBF. The other disadvantages of CBBF like poor cell protection, labor intensity, and dosage controversy, limited scope of export, expensive package and transport, very slow adaptation by the farmer community are some of the strongest problems which are being solved by the Liquid biofertilizers very effectively. Therefore, LBF are believed to be the best alternative for the conventional carrier based biofertilizers in the modern agriculture research community witnessing the enhanced crop yields, regaining soil health and sustainable global food production.

The Concept of Liquid Biofertilizers

"A preparation comprising requirements to preserve organisms and deliver them to the target regions to improve their biological activity (or) A consortium of microorganisms provided with suitablemedium to keep up their viability for certain period which aids in enhancing the biological activity of the target site". Liquid formulation is a budding technology in India and has very specific characteristics and uniqueness in its production methods. Liquid biofertilizers are the microbial preparations containing specific beneficial microorganisms which are capable of fixing or solubilizing or mobilizing plant nutrients by their biological activity [2].

They are broadly classified into three groups.

- Nitrogen Fixing Microbes (NFM)
- Phosphorus Solubilizing Microbes (PSM) and Phosphate Mobilizing Microbes (VAM) and
- Potash Mobilizing Microbes (Frateuria aurentia)

Among the three groups Phosphorus solubilizing microbes' usage is to the larger extent in the current agro market. The commercial production of Potash mobilizing bacteria has been developed recently. The authors presently concentrate on solubilization or mobilization of Micronutrients by exploiting the biological activity of microbes like

- Zinc and Sulphur Solubilizing Bacteria (Thiobacillus spp)
- Manganese solubilizer (Pencillium citrinum)

The biggest challenge of the biofertilizer production units is the viability of organisms up to field application. Some of the beneficial organisms are very effective in vitro, but may fail at some stage of cropping, even at the stage of harvesting. The common and potential causes of this demise are poor stability of the biofertilizer (by means of the viability) which may not be constant from prior storage to field application, very low potential ingredients reaching the active site of the plant practically and rapid degradation of the fertilizer by means of acclimatization problems. Liquid formulations of the beneficial organisms (LBFs) or bioinoculants are the promising inputs which can effectively combat the present agro adverse Scenario, resolving the low viability of the organisms, maximizing the efficacy of the inputs and satisfying the concept of cost-effectiveness.

TYPES OF INOCULUMS

Basically Liquid biofertilizers have been classified into, 1) Dry products which include dusts, granules and briquettes and, 2) Suspensions containing oil or water-based and emulsions. A wider range of formulations with additives are also available in the global agri market.

Dry Inoculum Products

These formulations comprise inoculums in semi dry condition. This includes dusts, granules and briquettes, based on particle or aggregate size. This group also comprises wettable powders formulated in dry powder. These powders require liquid or water as carrier just before the application to activate the inoculums. Dusts usually ranges from 5-20 mm in size and require low absorbance inert diluents as carriers. Low dusts particles, i.e. less than 10 mm size adhere best on the target site but are hazardous when inhaled. About 30% of the dry weight of dusts contains organisms as suspension. They are normally prepared by feeding the organism into an air stream for mixing with a blender. Particle size, bulk density and flow ability are extremely important.

Granules, Pellets, Capsules and Briquettes Inoculum

These include masses of different sizes which are usually called as Granules, pellets, capsules and briquettes. Granules are discrete masses 5-10 mm3 size, pellets are of size >10mm3 , and briquettes are large blocks up to several cubic centimeters; just like dusts, these products also contain inert carriers such as clay minerals, starch polymers and ground plant residues holding the organisms. Absorption (more important for formulating slurries of organisms), hardness, bulk density and product disburse rate in water etc., are the major factors taken into consideration in selecting a suitable carrier. Soft carriers e.g. Bentonite, disburse quickly to release the organism. Various materials are used to coat the product to slow down or control the rate of release, which also varies with the unit size. The usual concentration of organisms in granules is 20- 30%. There are three types of granular formulations in general: 1) the organisms attached to the outer surface of a granular carrier by a sticker in a rotating drum. 2) The organisms incorporated into a paste or powdered carrier which sets as a matrix size being controlled by passing the product through a sieve. 3) The organisms sprayed onto a rotating granular carrier without a sticker. When the carrier forms a protective coat around a core aggregate of organisms, the unit is termed as a capsule.

Wettable Powder Inoculums

These formulations comprise charcoal, lignite, vermiculite powders which are predominated among all commercial products. These are blended with 3% gum to make them stable on the shelf during storage and stick readily to the seeds.

The Issue of Formulations

Liquid formulations are spray able propagules of the microbial agents suspended in suitable liquid medium (or) A consortium of microorganisms provided with suitable medium to keep up their viability for certain period which aids in enhancing the biological activity of the target site. Basic characteristics of formulation

- To stabilize the organism during production distribution and storage.
- To deliver easily to the field in the most appropriate time.
- To protect the microorganism from adverse environmental factors at the target site (field), thereby increasing persistence.
- To enhance activity of the organism at the target site by increasing its viability, reproduction, contact and interaction with the target crops.

Although, India is considered to be the largest biofertilizers producer in the world, the creditability of biofertilizers is not up to the mark among farmer community. This is an alternative technology developed in the recent times resolving the adverse effects of agrochemicals and enhancing the activity of carrier based biofertilizers and biocontrol agents. This is not the usual broth culture from the fermenter or a water suspension of the carrier based biofertilizers. Liquid biofertilizers are special liquid formulation containing not only the desired beneficial microorganisms and their biological secretions, but also special cell Protestants or substances that encourage the formation of dormant spores or cysts for longer shelf life and tolerance to adverse conditions

Dormant Aqueous Suspensions

Dormant technology is still being used by several current commercial producers. Generally, they are using growth suppressants, contaminants suppressant like sodium azide, sodium benzoate, butanol, acetone, fungicides, and insecticides etc. for the long term viability. The Rhizobium dormant liquid formulations when used after 8 months reduce the size of nodules and effect nitrogen fixation process. This is due to the need of long duration of activation time. In case of Azospirillum, Azotobacter, PSM, it is observed that these bacteria crossed extreme dormant stage. Therefore, when applied in crops they take prolonged reactivation time. This long time is not desirable for short duration crops. Resolving this disadvantage a new technique was adopted in liquid formulations process involving the arrest of bacteria without preservatives.

Dormant Oil Suspension

Oil serves as the perfect medium to supply the inoculants in viable condition. Microorganisms can be suspended in oil at high concentration in various degrees of dehydration and remain viable. This formulation delivers organisms in a physiologically dormant state and does not encourage the growth of contaminants during storage. The bacteria/fungi have been successfully dried by continuous aeration as a suspension in oil to provide inoculants with shelf lives of several years. Viscosity roughly equals the setting rate of the particles. This is achieved by the use of colloidal clays, polysaccharide gums, starch, cellulose or synthetic polymers.

FACTORS AFFECTING LBF

Temperature

Temperature is important for the shelf life of microbial products and it can affect their activity pre or post microbial application. Temperature optima and limits vary with the microorganism. Colonization proceeds at field temperatures in the cropping season, but slows at lower temperature. Strains used in liquid formulation normally grow at 370°C [2] and able to tolerate temperature up to 450°C for two year or more. Whereas, solid base shelf life is hardly up to 3 months as rise in temperature beyond 35°C and start rapid decline of organisms.

Acclimatization

Effect of environment on bio fertilizers has been studied by many workers. It is reported that it has negative effect on different strains, but all the research was done on the carrier base biofertilizers. It is observed that the efficiency of liquid is almost same in all environments, but efficiency may reduce 20-25% in different climatic conditions in case of solid base. Normally in RCOF liquid formulation of an organism remain active for a long time after application, ideally throughout the period of the crop (2 years to the maximum), or in soil throughout the crop cycle. Temperature humidity, leaf surface exudates and competitors, etc are the critical factors which inactivates the microbes. They may be lost physically from the target location by the action of wind, rain or leaching depending on the fact why and where the product is used.

Effect of Humectants

The effect of moisture content on the storage, stability and activity is up to the greater extent. Some organisms may need moisture for their activity. In case of Carrier based inoculums the organism gets stressed due to number of reasons when carrier become dry during transport and storage. Bacteria needs wet plant surface in order to establish them and show its activity. These conditions can be fulfilled only by liquid formulation as they contain humectants. The level of direct effect of humidity on spore forming bacteria in liquid forms is relatively very low.

Sunlight Intensity

Generally Microbes are sensitive to temperature. The most harmful rays UVB (280-320 nm) and UVA (320-400 nm) reaching the earth's surface have direct effect on microbes. Of course some are less damaging where as some have the maximum effects. However, microbes are sensitive to the wave-lengths outside this range. To counter the harmful effects of high temperature sunscreens are added to a formulation. Sunscreens act by reflecting and scattering physically or by absorbing radiation selectively, converting short wavelengths to harmless longer ones. However, no such types of sunscreens are available in solid base to resist the effect of sunlight.

Effect of pH

The pH of a product plays a vital role in liquid inoculum preparation. It must be stabilized within certain ranges. The organisms are inactivated at the extreme high and low temperatures; therefore, a buffer is maintained by adding some additives which render the better shelf life in liquid. Maintenance of optimal pH improves shelf-life of some of microorganisms like Azospirillum, Azotobacter, Phosphorus Solubilizing bacteria (PSM), Potash Mobilizing Bacteria (KMB),

Frateuria aurentia [2].

Stabilization of Liquid Inoculums

The longest period in the life of a product elapses during storage, i.e. the time between manufacture and eventual use in the field. This period can range from several weeks to years. Organisms are required to remain viable during storage, with minimum loss of potency/ activity and without loss or breakdown of the desired formulation properties. The shelf life of liquid inoculum vs. carrier based inoculum has been shown in Table 1. The liquid formulations are added with suitable additives for improving stability which promote growth throughout the plant lifecycle by maintaining proper storage conditions prior to application by appropriate processing after production. Stability of the product is improved by resolving various physical and contamination problems like adhesion to the target site, activation time and the purity of the mother culture used for the subsequent preparation of the product respectively.

Table 1: Liquid Inoculum Vs Carrier based Inoculum -Shelf life; Source: Chandra et. al. 2004, Biofertilizer vision 2004; L-Liquid (Liquid formulation without preservative was used); S-Solid (lignite carrier mesh size 150 micron, pre-sterilized in autoclave used)

Months											
Inoculum	0	2	4	6	8	10	12	14	16	18	20	24
Azosp (L)	10.0	10.0	10.0	10.0	9.0	9.0	9.0	9.0	8.0	8.0	8.0	8.0
Azosp (S)	9.0	8.0	7.0	5.0	Nil	Nil	Nil	Nil	Nil	Nil	Nil	Nil
Azoto (L)	9.0	9.0	9.0	9.0	9.0	8.0	8.0	8.0	8.0	8.0	7.0	7.0
Azoto (S)	8.0	8.0	7.0	6.0	Nil	Nil	Nil	Nil	Nil	Nil	Nil	Nil
P.S.M (L)	9.0	9.0	9.0	9.0	9.0	9.0	9.0	9.0	9.0	9.0	9.0	9.0
P.S.M (S)	9.0	8.0	8.0	8.0	7.0	Nil	Nil	Nil	Nil	Nil	Nil	Nil
K.M.B (L)	9.0	9.0	9.0	9.0	9.0	9.0	9.0	8.0	8.0	8.0	8.0	8.0
K.M.B (S)	8.0	8 .0	8.0	7.0	5.0	Nil	Nil	Nil	Nil	Nil	Nil	Nil
Rhizo (L)	9.0	9.0	9.0	9.0	9.0	9.0	8.0	7.0	5.0	Nil	Nil	Nil
Rhizo (S)	9.0	8.0	8.0	7.0	Nil	Nil	Nil	Nil	Nil	Nil	Nil	Nil

HANDLING AND APPLICATION OF LIQUID INOCULUMS

Liquid inoculums ensure the facility of easy handling and application. This is exemplified by the addition of thickeners or suspenders to suspensions which aids in maintaining even distribution of the organism over the plant. Liquid prevents the clumping of the organism and ensure its ready resuspension after prolonged storage, whereas uniformity is not maintained by dusts and other wettable powders. It is evident that the population count come down to 105 in the duration of six months at room temperature in the case of Azospirillum while it survives in liquid up to 2 years and maintains population up to 108/ ml. Similarly the shelf life of Azotobacter, KMB and Rhizobium in solid state is up to 6 months and PSM can survive up to 8 months but liquid formulations of Azotobacter, PSM and KMB is up to 2 years followed by Rhizobium for 14 months

TYPICAL CHARACTERISTICS OF DIFFERENT LIQUID BIOFERTILIZERS

Rhizobium

Rhizobium is the classical example of symbiotic nitrogen fixing bacteria. The bacteria intact the legume root and form root nodules (nodular symbiosis) within which s is reduced to ammonia that is readily utilized by the plant to produce valuable proteins, vitamins and other nitrogen containing compounds. It has been estimated that 40-250 kg N/ha/year could be fixed by the microbial activities of Rhizobium for various legume crops. When the plants die the fixed N2 is released, making it available to other plants and this helps in fertilizing the soil. Majority of the legumes have this type of association, except a few genera (e.g., Styphnolobium) [1].The density of nodules occupied, dry weight of nodules, dry weight of plant and the grain yield per plant influenced by the multi strain inoculants was highly promising.

However, it is evident from the research that Liquid biofetilizers especially liquid rhizobia with 2% PVP were found to maintain higher CFU count and was very effective for enhancing the soya bean yield particularly in US (Hawaii Under NifTal), Brazil, Europe (e.g., Spain, Argentina) and Asia (China, Vietnam) [4].

Physical Features of Liquid Rhizobium

- Dull white in colour
- No bad smell
- No foam formation, pH 6.8 to 7.5

Use of Fungicides/Pesticides/Herbicides with Liquid Rhizobium Inoculum: The effect of chemicals on nodulation and nitrogen fixing Rhizobium is negligible. Seed dressing fungicides such as Thiram and mercury have some adverse effect on the viability of Rhizobium, which can be overcome by doubling the doze of Rhizobium culture. Studies suggest that Rhizobium strain MRL3 may be exploited as a bioinoculant to augment the efficiency of lentil exposed to insecticide-stressed soils. [5]. On the other hand fungicides such as Dithane M-45 and Bevistin do not have any adverse effect on Rhizobium. The minimum tolerance doses of five different fungicides, Corbandazium, Dithane M-45, Dithane Z-78, Topsin and Ceresin were tested for Bradyrhizobium japonicum and Rhizobium leguminosarum locally isolated from Glycine max and Pisum sativum. Amongst these fungicides, ceresin was found to be more toxic to Bradyrhizobium japonicum and Rhizobium leguminosarum, since the minimum tolerance dilution is 107, whereas minimum dilution of Dithane Z-78 for soybean is 103; 104 for pea; Dithane M-45 for soybean is 104; pea is 105; Topsin for pea is106; soybean is 107 and Corbandazium for pea and soybean being 106.

Dithane Z-78, Topsin and Ceresin were tested for Bradyrhizobium japonicum and Rhizobium leguminosarum locally isolated from Glycine max and Pisum sativum. Amongst these fungicides, ceresin was found to be more toxic to Bradyrhizobium japonicum and Rhizobium leguminosarum, since the minimum tolerance dilution is 107, whereas minimum dilution of Dithane Z-78 for soybean is 103; 104 for pea; Dithane M-45 for soybean is 104; pea is 105; Topsin for pea is106; soybean is 107 and Corbandazium for pea and soybean being 106.

Use of Sticky Materials for Seed Treatment: Generally sticky materials are used for the easy adhesion of the inoculums onto the target plant parts and subsequently increase bacterial population on the seeds for maximum nitrogen fixation. The germination and nodulation tests were performed on the pellated seeds and results showed that although the sugar and molasses initially bound less inoculant and lime to the seed, the number of surviving rhizobia was similar to that obtained by the Gum arabic treatment after storage at 27.50°C for five days for fast growing rhizobia. The inoculated pellated seeds of slow growing rhizobia showed promising results in germination and nodulation in acidic soil.

Azospirillum

Azospirillum belongs to bacteria and is known to fix considerable quantity of nitrogen in range of 20-40 kg N/ha in the rhizosphere in non-leguminous plants such as cereals, millets, oilseeds, cotton etc. Azospirillum is considered as the efficient biofertilizers because of its ability of inducing abundant roots in several plants like rice, millets and oilseeds even in upland conditions. An estimated amount of 25-30% chemical nitrogen fertilizer can be saved by the appropriate use of Azospirillum inoculants. The genus Azospirillum has three species viz. A. lipoferum, A. brasilense and A. amazonense. These species have been commercially exploited for the use of nitrogen supplying biofertilizer. One of the characteristics of Azospirillum is its ability to reduce nitrate and denitrify. Table 2 presents occurrence of Azospirillum and their nitrogen fixing capacity in different Plants. Both the species A. lipoferum and A. barasileznse may comprise strains which may denitrify actively or weakly or reduce nitrate to nitrite. Therefore, for inoculation preparation, it would be necessary to select strains which do not possess these characteristics. Azospirillum lipoferum present in the roots of some tropical forage grasses such as Digitaria, Panicum, Brachiaria, Maize, Sorghum, Wheat and Rye.

Table 2: Occurrence of Azospirillum and their nitrogen fixing capacity in different Plants

Plant	N2 fixed mglg of substrate
Sorghum bicolor (Sorghum)	20
Zea mays (Maize)	20
Panicum sp.	24
Cynodon dactylon	36
Setaria sp.	12
Amamanthus spinosa	16
Oryza saliva (Paddy)	28

Physical Features of Liquid Azospirillum: The color of the liquid may be blue or dull white. Bad odors confirm improper liquid may be broth. Production of yellow gummy colour materials confirms the quality product. Acidic pH always confirms no Azospirillum bacteria present in liquid.

Production of Growth Hormones: Azospirillum cultures synthesize considerable amount of biologically active substances like vitamins, nicotinic acid, indole acetic acid, gibberellins. All these hormones/ chemicals help plants in better germination, early emergence, and better root development.

Field Applications of Liquid Azospirillum

- Stimulates growth and produces green color characteristics of a healthy plant.
- Aids utilization of potash, phosphorus and other nutrients
- Encourages plumpness and succulence of fruits and grains and increases protein percentage.

Inoculants of Azospirillum have been proved beneficial in a variety of crops including cereals, forage and other crops. The crop response however, varies greatly with types of crop and variety, location, season, soil fertility level; native microorganism interaction etc., the ability to fix atmospheric nitrogen and to produce photo hormones determines the crop response and yield. A series of experiments have been conducted on a range of crop varieties to evaluate the crop response to Azospirillum inoculants and a wide variation in production up to 40% have been observed. Nonfunctioning of Azospirillum in the fields can be identified by yellowish green color of leaves which indicates very poor or no nitrogen fixation which in turn results in arresting the growth promotion.

Azotobacter

Azotobacter belongs to the genus diazotrophic bacteria which is a free-living organism whose resting stage is a cyst. It is abundantly found in neutral to alkaline soils, in aquatic environments, and on some plants. Azotobacter is capable of performing several metabolic activities, including atmospheric nitrogen fixation by conversion to ammonia. Azotobacter spp. has the highest metabolic rate of any organisms. It serves as potential biofertilizer for all non-leguminous plants especially rice, cotton, vegetables etc. Azotobacter population is high in rhizosphere region and the rhizoplane does not contain Azotobacter cells. The lack of organic matter in the soil is a limiting factor in the proliferation of Azotobacter in the soil.

Liquid Azotobacter in Tissue Culture:

- The performance of Azotobacter liquid inoculants was comparatively better than all the treatments in 10% MS medium followed by Azospirillum.
- The performance of Azotobacter liquid inoculants was comparatively better than all the treatments followed by Azospirillum for the growth of polybag sugarcane seedlings.

Liquid Azotobacter as Biocontrol Agent: Azotobacter have been found to produce some effective antifungal substance which inhibits the growth of some soil fungi like Aspergillus, Fusarium, Curvularia, Alternaria, Helminthosporium, Fusarium etc.

Acetobacter

Acetobacter, a sacharophilic bacteria associated with sugarcane, sweet potato and sweet sorghum, coffee plants is acid-producing bacteria which fix atmospheric nitrogen. In fact, the A. diazotrophicussugarcane relationship is beneficial symbiotic relationship between grasses and bacteria which fixes about 30 kg/ha/year of nitrogen. This bacterium is being exploiting and commercialized extensively for sugarcane crop reducing the primary dependency in the chemical Nitrogen fertilizer. It is known to increase cane yield 10-20 ton/acre and sugar content by 10-15 percent.

Liquid Acetobacter Diazotrophicus in Sugarcane: Application of Azospirillum and Phosphobacteria for the cash crop sugarcane has been a regular practice for the past few years which is saving nearly 20% chemical nitrogen and phosphate applications in South India. It is evident from various studies that Acetobacter diazotrophicus present in sugarcane stem, leaves, soils have a capacity to fix up to 300 kg of nitrogen. This bacterium first reported in Brazil where sugarcane is cultivated in very poor sub-soil fields which are fertilized with phosphate, potassium and micro elements alone could produce a yield for three consecutive recorded harvests of 182 to 244 tons per hectare, without the use of any chemical nitrogen fertilizer. This manifests the assumption that active nitrogen fixing bacteria must be associated within the plant regularly. An extensive research work has been conducted by the scientists of Brazil and Australia on bacteria like Acetobacter diazotrophicus, Herbaspirillum seropedicae and Herbaspirillum rubisubalbicans in leaves, stems and roots and revealed the fact that they are largely responsible for atmospheric nitrogen N-fixation.

Acetobacter Pasteurianus for Sulfur: The physical structure of sulfur consists of several amino acids, including methionine and cystine, which are essential components of plant and animal proteins. Sulfur is important in synthesis of nitrogen in plants; Sulfur acts more like nitrogen than any other essential plant nutrients. Indeed, sulfur deficiency symptoms in plants closely resemble nitrogen deficiency symptoms. Moreover, leaching losses of sulfur from the soil are very much like those of nitrogen caused by percolation waters through lighttextured sandy and salty soils. Thus, there is need for annual supplies of sulfur to meet the requirements of growing plants. Although, we are accustomed to use sulfur in the form of fertilizers but number of such farmers are very less, due to its cost factors etc. It was estimated that about 70-90% of total surface soil sulfur is found in organic matters. The remainder of the sulfur is present as sulfides and as soluble and insoluble sulfates. Usually, only 5-10% of the sulfates are found in the surface furrow slice of humid region, cultivated soils. In humid region soils, most of the sulfates are of 12-14 inches deep and are associated with oxides of iron and aluminum, especially in acid soils having Kalononitic type clay. These bacteria help in converting this nonusable form to usable form. It was experienced that use of 200 ml/acre bacteria influenced in level of sulfur in crops like vegetable, cabbage, turnip, onion, cotton, fruits, oilseeds, spices etc.

Phosphorus Solubilizing Microorganisms

Microbial mineralization, solubilization and mobilization are the most important aspects of the phosphorus cycle, besides chemical fixation of phosphorus in soil. The enzymatic activity of microorganisms is responsible for the mineralization of organic phosphorus which is left over in the soil after harvesting, or added as plant or animal residues to soil [6,7].

Phosphorus solubilizing bacteria and fungi play a vital role in persuading the insoluble phosphatic compound such as rock phosphate, bone meal and basic slag and particularly the chemically fixed soil phosphorus into available form. These special types of microorganisms are known as Phosphate Solubilizing Microorganisms (PSM) which includes different groups of microorganisms such as bacteria and fungi that convert insoluble phosphatic compounds and fixed chemical fertilizers into soluble form. The species of Pseudomonas, Micrococcus, Bacillus, Flavobacterium, Penicillium, Fusarium, Sclerotium, Aspergillus and some other are considered as active in biophosphorus conversion.

These bacteria and fungi can be grown in the media containing $Ca_3(PO_4)_2$, $FePO_4$, $AlPO_4$, apatite, bone meal, rock phosphate or similar insoluble phosphate compounds are the sole source of phosphate. Such organisms not only assimilate phosphorus but also trigger the release of an excess amount of soluble phosphate than their actual requirements. Several rock phosphate dissolving bacteria, fungi, yeasts and actinomyctetes were isolated from soil samples collected from rock phosphate deposits and rhizosphere soils of different legume crops. Some of the isolates solubilized and made available phosphorus to crop from rock phospshate and many isolates solubilized a very high quantity of tricalcium phosphate.

The most efficient bacterial isolates were identified as Pseudomonas striata, Pseudomonas rathonis, Bacillus polymyxa, B. megatherium. Apart from these, fungal isolates such as Aspergillus awamori, Penicillium digitatum, Aspergillus niger, Schwanniomyces occidentalis are identified as the potential phosphorous solubilizers. These efficient microorganisms have shown consistency in their capability to solubilize chemically fixed soil phosphorus and rock phosphate from different sources. PSM were found to mineralize organic phosphorus into soluble form by their enzymatic activity and save $P_2 O_5$ up to 30-50 kg/ha.

Physical Features of Liquid PSM: High turbidity is observed in the appearance of the solution. The color may be creamy mostly and it varies along with the strain. The odor may be pleasant but sometimes mild. The pH is 4.5 or buffer, but buffer with 6.5-7.0 ensures more than 2 years viability.

Effect of Liquid PSM under Field Conditions: PSM encourages early root development, produce organic acids like malic, succinic, fumaric, citric, tartaric and alpha ketoglutaric acid which hastens the maturity and thereby increases the ratio of grain to straw as well as the total yield. Improves others beneficial microbial compatibility with the plants. Helps in rapid cell development in plants and consequently enhance diseases resistance towards pathogens. Increase micro nutrient content in soil like Mn, Mg, Fe, Mo, B, Zn, Cu etc., and make them available to the plant parts; stimulates formation of fats, convertible starches and healthy seeds.

Inoculants of phosphate solubilizing bacteria as fertilizer increases P uptake by the plant and enhance crop yield. There are many strains of different genera which promote the phosphate solublization among which, strains from the genera Pseudomonas, Bacillus and Rhizobium are the most powerful phosphate solubilizers. Phosphate solubilization by the organisms mainly involves the production of organic acids which is the principle mechanism solubilizarion, and the organic phosphorous in the soil is mineralized by acid phosphates which is a vital step. Cloning of several phosphatase encoding genes which are isolated is done and characterized for the better phosphate solubilization. Genetic manipulation of phosphate-solubilizing bacteria followed by their expression in selected rhizobacterial strains is the interesting approach among the scientific community for the optimum phosphate solubilization [8]. The PSB has the greatest potential of utilization of India's abundant deposits of rock phosphates, much of which is not enriched [9].

Phosphate Mobilizing Microorganisms

Compared with the other major nutrients, phosphorous is the least mobile and available to plants in most of the soils. In spite of its abundance in the most of the soils both in organic and inorganic forms, phosphorous is prime limiting factor for plant growth. Moreover plant species, nutritional status of the soil and ambient soil conditions determine the bioavailability of soil inorganic phosphorus. To circumvent the phosphorus deficiency I the soil phosphate solubilizing microorganisms (PSM) could play an important role in supplying phosphate to plants in a more suitable and eco friendly manner [10].

Several soil fungi are capable to mobilizing/making it available to the plant from the immobile form of phosphorous by its hyphal structures and hence called as phosphate mobilizing microorganisms. These soil microbes have mutualistic association with all crop plants except family Brasicaceae. Besides phosphorus this fungus also mobilizes zinc and sulphur. The literal meaning of mycorrhizae is "fungus root" which denote the non pathogenic association between certain soil fungi and plant roots. In general two typical types of mycorrhizal fungi are found in associations with plant roots namely ectomycorrhiza, endomycorrhiza and ectendomycorrhiza. Mycorrhizae increase the surface area of the plant root system for better absorption of nutrients from soil especially in phosphorus deficient soils. The endomycorrhiza, also known as Arbuscular mycorrhiza, aids in transfer of nutrients from soil into root system through specialized structures known as vesicles and arbuscules. The potential role of mycorrhiza in mobilizing & phosphate uptake and plant growth is being widely recognized in the recent times and Arbuscular mycorrhizas (AM) are the most common type which is being highly exploited. Advancement in biotechnology has promoted experimentation in artificial inoculation of plants with AM. Trials have mainly involved broadcasting and raking-in inocula over plant-growing substrates rather than application of inocula to seed before sowing. Since large-scale production of AM in axenic culture is not yet attained, inocula have been produced in pot cultures or in small field plots on plants grown under carefully controlled conditions, to avoid contamination by plant pathogens. Such inocula have comprised infected roots or spores and hyphae trapped in soil, peat and clay carriers. Introduction of AM into field on a large scale has tended to use inoculum at excessively high and impractical rates in order to guarantee rapid infection by the introduced fungi. Application of inoculum directly to seed may reduce the amounts of inoculums required. Experimental inoculation has involved coating spores, mycelium and infected root fragments, wet-sieved from soil, onto seeds using methyl cellulose as sticker.

According to the studies of root nodulation as well as N and P, uptake is improved by green gram plant on the application of favorably interacting rhizospheric microorganisms as the inoculants and hence, yield is increased in the phosphorous deficient soils [11]. Bradyrhizobium sp. (vigna) and phosphate solubilizing microorganisms (PSM), Pseudomonas striata or Penicillium variable have shown a significant effect in enhancing the nutrient uptake and plant yield. Combined inoculation treatment of AM fungus, Glomus fasciculatum the with Bradyrhizobium sp. (vigna) + P. striata further augmented the nutrient uptake and plant yield [10].

BENEFITS AND CONSTRAINS

As the fungal hyphae facilitate increased surface area of absorption of nutrients like P, Zn, Cu etc., up to an extent of 8 cm the biochemical reactions in plant cycle are not interrupted promoting the plant growth.

Moreover, it is rather easy to inoculate on crops which are normally grown on nursery beds or root trainers or poly bags because of the difficulty in mass producing AM fungi. Nearly 25 to 50% of phosphatic fertilizers can be saved on a year with the inoculation of efficient AM fungi.

In spite of many advantages, AMF have some constraints which are causes for their limited usage. As it is an obligate symbiotic organism it need a host to survive, and it cannot be cultured in synthetic media in laboratory for large scale production. Variations and specifications in plant genotypes and soil nature make it difficult to use appropriate mycorrhizal strains for the phosphorous mobilization. Slow growth and development of mycorrhizal fungi association with the plant roots and poor resistance to the fungicides are some of the other potential constrains which are to be overcome by the rigorous field research with different approaches and optimizations.

POTASH MOBILIZING BACTERIA (KMB)

In addition to the regular biofertilizers like Rhizobium, Azospirillum, Azotobacter, Acetobacter and Phosphate solubilizing/mobilizing and microbes to meet N&P nutrition in plants, microbes responsible for Potassium mobilization has been isolated and developed from Banana rhizosphere which is having the ability to mobilize the elementary or mixture of potassium which can be easily absorbed by plants. It is estimated that 50 to 60% of potash chemical fertilizers usage can be reduced by using Frateuria aurentia, a new bacterial species (species conformation by IMTECH Chandigarh) as a bioinoculant. This new bacteria belonging to the family Pseudomonaceae have the extra ability to mobilize K in almost all types of soils especially, low K content soils, soils of pH 5-11 and it survives in the temperature up to 42°C. This potash mobilizing biofertilizers can be applied in combination with Rhizobium, Azospirilum, Azotobacter, Acetob acter, PSM etc. Potash mobilizing bacterial based product containing Frateuria aurentia producing plant growth promoting substances which offers plant a multifaceted benefits in terms of growth, by mobilizing potash and making it available to crops. It also enhances the efficiency of chemical fertilizer [12].

EFFECT OF LIQUID POTASSIUM MOBILIZER WITH AM

Optimization studies on the Liquid Potassium mobilize revealed following facts. Arbuscular mycorrhae (soil base) and Potassium mobilizing bacteria, individually, as well as in combination with organic compost and phosphocompost shown better results in capsicum, Hungarian yellow variety when compared to the controls (plain soil). Potassium mobilizing bacteria (Frateuria aurentia) enriched phosphocompost shows the highest results in growth parameters next to all the biofertilizers applied together. This may be because of significant increase in mobilization and uptake of K leading to release of growth hormones by the K-mobilizer.

CONSTRAINTS

Even though there is a considerable refinement in the field of biofertilizers for the past 3-4 decades the biofertilizer industry has been facing some technological constraints, some being resolvable the rest need strong research & development activities.

Crisis of Efficient Strains

Unavailability of potential regional strains is one of the major reasons. The specificity and competitive ability of the strain is the key point on which the efficacies of the organism relay with respect to the hosting soil and plant variety. The ability to fix Nitrogen and survival capacity are the limiting factors of the Liquid as well as the carrier based biofertilizers [13].

Possible Genotypic Changes

During the production of the fertilizers the organism may get interacted with other organisms which may leads to change in basic character of the organisms. Apart from this during fermentation the strains may undergo mutations which may alter the efficacy and viability leading to the economical loss.

Lack of Awareness

In spite of many ongoing projects on the development of biofertilizers, proper attention towards the technology is still needed in order to manifest the results at field level. Communication gap and miscommunication of the farmers by some commercial producers also is a considerable constraint. Lack of storage and marketing facilities, consistent awareness in the farmers by conducting community programs is still welcome [14]. As the production and distribution of biofertilizers is seasonal, the commercial production units may suffer from lack of demand. Moreover the establishment of microorganisms resisting the antagonistic activity of the native microbes is always challenging. Soil nature, temperature, humidity, ph, local insect population, etc., are some other acclimatization problems [15].

- FUTURE STRATEGIES
- Identification and characterization of potential organisms (unexplored) and their effective exploitation in the field of Agriculture is urgently needed.
- Selection and application of suitable bioinoculants with respect to soil nature, agro climatic condition, crop variety under proper agriculture practices is needed.
- Genotypic study of the strains and molecular characterization of the plant parts is necessary to understand the plant mechanisms.
- Study of soil texture and compatible studies with respect to microbial interventions.
- Exploring the novel soil bacteria and maintaining the genomic libraries for future exploitation.
- Identifying, studying the population genetics and preserving the useful endangered species and by advanced bio molecular techniques and bioinformatics tools.
- Suitable combinations of microbial formulations (liquid microbial consortium) with optimized field results are preferable for the sustainable production.
- Soil analysis, crop rotation, organic manure usage, maintenance of proper moisture content, regular sterilization practices are emphasized which are necessary to maximize the biofertilizer efficacy

- Global standards in the research & development should be maintained during the production and storage of the formulations.
- Development of new strains with enhanced capabilities by genetic engineering techniques and rDNA technology is needed to maintain an ecofriendly & sustainable agriculture.
- Constructive awareness and technical support by microbiologists and agricultural professionals must be provided to the Agrarians.

CONCLUSIONS

Human has been relaying on agriculture throughout the evolution and would be depending on it forever. The impact of globalization, transition in the technology is creating a negative shade on agriculture in the developing countries like India and declined the percentage of the farming community drastically resulting in raising demand for food commodities. Growing global population demanding the safe and sufficient food for the survival Soil heath has become the greatest assertion for the scientific community in this ever growing polluted globe. Uncertainty in the agro climatic conditions (edapho-climatic factors), monsoon failures by the priceless human activities, lack of proper awareness among the farming community are the direct causes for the agriculture failure in the developing countries like India. The rising demand for the fields like food processing, packing industries, ready to eat foods etc., witness the demand for raw materials in agriculture sector. Crises of agricultural land day by day, vertical increase in the cost of agriculture input technologies are leading to transitions in farming community. In such an agro critical scenario, a multifaced solution for different constraints in agro industry is necessary. It is evident that Biofertilizer technology has inaugurated a new era in biological input technology and recorded a tremendous raise in the annual agriculture production particularly in the past two decades. To combat the threat of global food crises the alternative technologies in the agriculture like liquid biofertilizers are obligatory. Liquid biofertilizer of course have the capacity to replace the traditional chemical fertilizers & carrier based biofertilizers and plays a major role in restoring the soil health, but a lot of measures in terms of technology, government support, subsidies, and constructive awareness by well trained technicians among the agrarians are emphasized.

REFERENCES

1. Mia MAB, Shamsuddin ZH (2010) Rhizobium as a crop enhancer and biofertilizer for increased cereal production. African Journal of Biotechnology 9: 6001-6009.
2. Chandra K, Greep S, Ravindranath P, Srivathsa RSH. Liquid Biofertilizers, Ministry of agriculture department of agriculture & co-operation, Government of India.
3. Mahdi SS, Hassan GI, Samoon SA, Rather HA, Dar SA, et al. (2010) Biofertilizers in Organic Agriculture. J Phytol 2: 42-54.
4. Sharma MP, Sharma SK, Dwivedi A (2010) Liquid biofertilizer application in soybean and regulatory mechanisms. Agriculture Today 44.
5. Ahemad M, Khan MS (2011) Insecticide-tolerant and plant-growth-promoting Rhizobium improves the growth of lentil (Lens esculentus) in insecticidestressed soils. Pest Manag Sci 67: 423-429.
6. Bot A, Benites J (2005) The importance of soil organic matter. Natural resources management and environment department, FAO corporation document repository.
7. Eckert D. Efficient Fertilizer Use-Nitrogen.
8. Rodrı guez H, Fraga R (1999) Phosphate solubilizing bacteria and their role in plant growth promotion. Biotechnol Adv 17: 319-339.
9. Ghosh N. Promoting Bio-fertilizers in Indian Agriculture. Institute of Economic Growth University Enclave, India.
10. Khan MS, Zaidi A, Wani PA (2009) Role of phosphate solubilizing microorganisms in sustainable agriculture - A review. Biomedical and Life Sciences 551-570.
11. Zaidia A, Khan MS, Aamila M (2004) Bioassociative effect of rhizospheric microorganisms on growth, yield, and nutrient uptake of Greengram. J Plant Nutr 27: 601-612.
12. Patel BC (2011) Advance method of preparation of bacterial formulation using potash mobilizing bacteria that mobilize potash and make it available to crop plant. WIPO Patent Application WO/2011/154961.

13. Annual Report 2010-2011. Department of Agriculture and Cooperation Ministry of Agriculture Government of India.
14. Revellin C, Giraud JJ, Silva N, Wadoux P, Catroux G (2001) Effect of some granular insecticides currently used for the treatment of maize crops (Zea mays) on the survival of inoculated Azospirillum lipoferum. Pest Manag Sci 57: 1075-1080.
15. Ngampimol H, Kunathigan V (2008) The Study of Shelf Life for Liquid Biofertilizer from Vegetable Waste. Au J T 11: 204-208.

Chapter 10

Novel Techniques in the Production of Industrially Imperative Products

Sameera V

Biotechnology Department, REVA Institute of Science and Technology, Bangalore University, India

ABSTRACT

The products which are being developed by the biotechnology industry have huge implications for our future health and well-being. All these discoveries in current biotechnical research and its applications will have repercussions within the human history. In recent years, microbial fermentations have been revolutionized by the application of genetically-engineered microorganisms. Economic importance of bacteria describes mainly from the fact that they are extensively exploited by humans in a number of beneficial ways. Fermentation is one such phenomenon in which microorganisms like bacteria, fungi, algae, etc are utilized for production of novel products. Industries like, diary, food, textile, baking etc are most benefited by these fermented products. Present article mainly focuses on the recent techniques which are involved in enhanced production of imperative products like lactic acid, cellulase, Human interferon α, and exopolysaccharide, and advantages of fermentation technology.

INTRODUCTION

Microorganisms are capable of growing on a wide range of substrates and can produce a remarkable spectrum of products. The relatively recent advent of in vitro genetic manipulation has extended the range of products that may be produced by microorganisms and has provided new methods for increasing the yields of existing ones. The commercial exploitation of the biochemical diversity of microorganisms has resulted in the development of the fermentation industry and the techniques of genetic manipulation have given this well-established industry the opportunity to develop new processes and to improve existing ones. The term fermentation is derived from the Latin verb fervere, to boil, which describes the appearance of the action of yeast on extracts of fruit or malted grain during the production of alcoholic beverages. However, fermentation is interpreted differently by microbiologists and biochemists. To a microbiologist the word means any process for the production of a product by the mass culture of microorganisms. To a biochemist, however, the word means an energy-generating process in which organic compounds act as both electron donors and acceptors, that is, an anaerobic process where energy is produced without the participation of oxygen or other inorganic electron acceptors [1]. Nowadays there is an increase in interest in the development of more efficient and less time consuming methods to measure the presence of microorganisms, as well as their viability for bioprocess control and improvement [2]. Identification of specific microorganisms is very important for carrying out fermentation process [3]. The manufacture of industrially important products by the methods of modern biotechnology is separated into two stages: upstream processing during which proteins are produced by cells genetically engineered to contain the desired gene which will express the product of interest and downstream processing during which the produced products are isolated and purified [4] (Figure 1).

Upstream Processing

Today, upstream processing can also take place in whole animals or plants, making the animal or plant a genuine bioreactor.

E.g. Monoclonal antibody (Mab) production using mice.

Other protein pharmaceuticals have been produced in the milk of goats and sheep and the leaves of tobacco to mention a few instances of this sort of upstream production of proteins [4].

Upstream processing may involve steps like preparation of inoculums, scaling up, inoculating the inoculums in the culture media/reactor, assessment of qualitycontrol periodically during various steps of fermentation, controlling of pH, temp, etc. (Figure 2) [4].

Downstream Processing

In a biochemical process, the highest operating and equipment costs can often be found in the area of downstream processing [5]. Downstream processing mainly refers to the recovery and purification of biosynthetic products, particularly pharmaceuticals, from natural sources such as animal or plant tissue or fermentation broth, including the recycling of salvageable components and the proper treatment and disposal of waste. Downstream processing and analytical bioseparation both refer to the separation or purification of biological products, but at different scales of operation and for different purposes. Downstream processing implies manufacture of a purified product fit for a specific use, generally in marketable quantities, while analytical bioseparation refers to purification for the sole purpose of measuring a component or components of a mixture, and may deal with sample sizes as small as a single cell. A widely recognized heuristic for categorizing downstream processing operations divides them into four groups which are applied in order to bring a product from its natural state as a component of a tissue, cell or fermentation broth through progressive improvements in purity and concentration [6] (Figure 3).

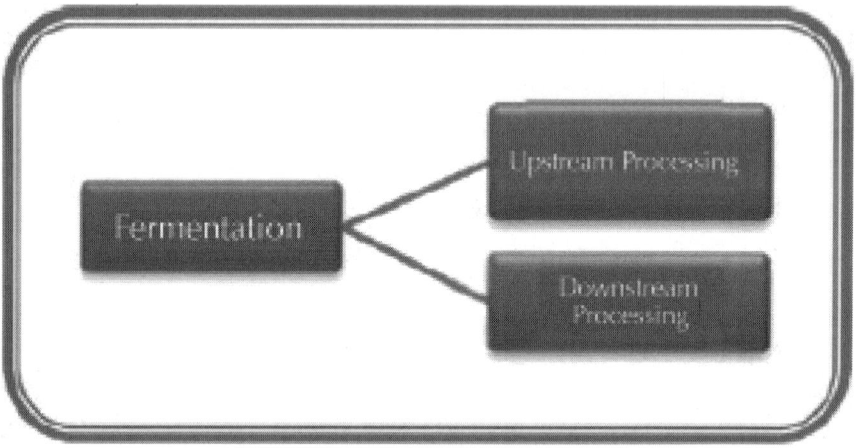

Figure 1: Classification of Fermentation.

Downstream processing steps mainly involve many techniques such as chromatography, flocculation, crystallization, etc. The rapid development of biotechnology and biomedicine requires more reliable and efficient separation technologies for the isolation and purification of biopolymers such as therapeutic proteins, antibodies, enzymes and nucleic acids [7].

Following are the methods which are used for purification of product [6].

- »» Filtration
- »» Centrifugation
- »» Sedimentation
- »» Precipitation
- »» Flocculation
- »» Adsorption
- »» Chromatography
- »» Crystallization
- »» Fractional distillation

Bioreactors like TOROCELL have also been developed by by Ravindranath Gandlur and Prasadarao Gandlur of Lablinks Biotech

Pvt. Ltd. recently for the cultivation of *Saccharomyces boulardii* as a probiotic supplement, *Escherichia coli* for plasmid DNA, *Streptomyces avidini* for producing Streptavidin, BHK 21 [8].

Applications of Fermentation

There is at present a world-wide need to develop the way we handle wastes and to produce energy in a sustainable manner. Taking the waste product of one process and using it as input or fuel for another process is one way to accomplish this by making intelligent use of resources [9]. One such way is fermentation, in which we can use wastes of sugar industry, (molasses) for the production of alcohol. Fermented food and beverages have been used worldwide since time immemorial [10]. Microbial fermentations may be classified into the following major group

- Production of microbial cells (biomass) as the product.
- Production of microbial metabolites.
- Production of microbial enzymes.
- Modification of a compound which is added to the fermentation the transformation processes.
- Production of recombinant products [1].

Figure 2: Steps in upstream processing.

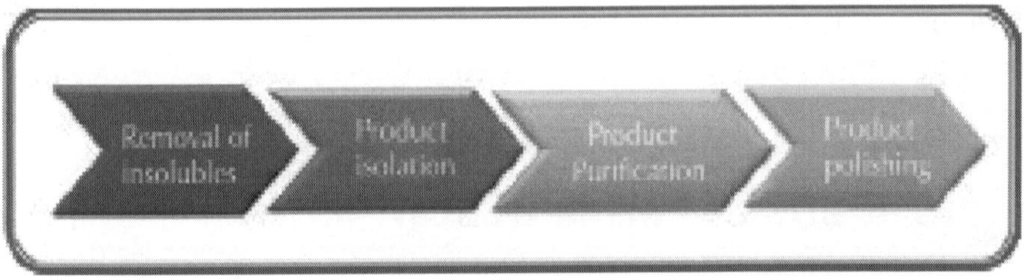

Figure 3: Steps involved in downstream processing.

Microbial Biomass

Microbial biomass is produced commercially as single cell protein (SCP) for human consumption or animal feed and as viable yeast cells for use in the baking industry.

Microbial Metabolites

The metabolism is defined as the sum of all the biochemical reactions carried out by an organism. Primary metabolic pathways converge into few end products while secondary metabolic pathways diverge into many products. Some of these metabolites which are industrially important can be produced by fermentation [11].

E.g. citric acid, alcohols (ethanol using *Saccharomyces cerevisiae* [12]) , etc

Microbial Enzymes

Enzymes are biocatalysts produced by living cells to bring about specific biochemical reactions in the cells. Each single strain of organism produces a large number of enzymes, which are hydrolyzing, oxidizing or reducing, and metabolic in nature.

E.g. Lipase [13,72], amylase, glucosidase, penicillinase, Fibrinolytic Enzymes [14], Asparaginase [15-17] etc

In the process of fermentation, selection of strains for the commercial production of specific enzymes is carried out, which have the capacity for producing highest amounts of the enzymes desired. Commercial enzymes are produced from strains of molds, bacteria, and yeasts [18].

Transformation Processes

In some cases of fermentation, microbial cells may be used to catalyse the conversion of a compound into a structurally similar, but financially more valuable, compound. Such fermentations are termed biotransformations, transformation processes, or bioconversions. The reactions that can be catalyzed comprise oxidation, dehydrogenation, hydroxylation, dehydration condensation, decarboxylation, deamination, amination, and isomerization [1].

Recombinant Products

Genes from higher organisms can be introduced into microbial cells for synthesizing foreign (or heterologous) proteins. Examples of the hosts include *Escherichia coli, Saccharomyces cerevisiae* and other yeasts as well as filamentous fungi such as *Aspergillus niger var. awamori*. Products produced in such genetically manipulated organisms include interferon, insulin [19], human serum albumin, factor VIII and factor IX, epidermal growth factor, bovine somatostatin and bovine chymosin.

E.g. Production of interferon [20], insulin, human serum albumin, factor VIII and factor IX, epidermal growth factor, bovine somatostatin and bovine chymosin [1], fusion proteins [21], production of antibiotics [22].

Although analysis of select compounds and metabolites has been used broadly to elucidate specific metabolic pathways, it is now more widely being applied to cataloguing all (or most of) the biochemical coupled with an organism's metabolism (or metabolome) [23]. Analysis of the biochemicals associated with an organism plays a crucial role in selection of strains in fermentation.

A single vertical rotating immobilized cell reactor (VRICR) with the bacterium *R. Erythropolis* as a biocatalyst, was developed and used for investigation of biodesulfurization procedure with its two consecutive stages of cell growth and desulfurization activity by fermentation [24].

Other Applications

- Production of fermented food [13, 25].
- Production of refreshing beverages by the help of "Tea fungus" (kombucha) [26]
- Production of biofuels [27] like ethanol, biogas [28], biomethanol [3]
- Production of pharmacological compounds like salidrosides [29]
- Production of human growth hormones [30,31]
- Production of Bioplastics [23]
- Lipases produced through fermentation are of great importance in food industry in development of flavour esters [32]
- Grapes and apples have been widely applied to ferment beverages [33] the use of other fruits, such as orange[34], cacao [35], mango [36] , gabiroba [37] , caja [38], kiwi [39], and in the production of wine has been recently demonstrated [40].
- Production of mycoprotein [41]
- Production of xylitol [42].
- Ethanol is considered to be the next generation transportation fuel with the most potential, and significant quantities of ethanol are at present being produced from corn and sugar cane by fermentation process [43].
- Production of Pectinmethylesterase by solid state fermentation [44].

In this article, we are going to discuss the fermentation of a range of products which comprise of the following

- Enzyme production
- Polysaccharides production
- Organic acid production
- Recombinant protein production

Production of Enzymes

Enzymes are among the most significant products obtained for human needs through microbial sources. They are catalysts which speed up a reaction without being used up in the reaction. A large number of industrial processes in the areas of industrial, environmental and food biotechnology utilize enzymes at some stage or the other. Current developments in biotechnology are yielding new applications for enzymes [45]. The starting point for enzyme production is a vial of a selected strain of microorganisms. They will be nurtured and fed until they multiply many thousand times. Then the desired end-product is recovered from the fermentation broth and sold as a standardized product [46].

Different kinds of processes can be used for enzyme production such as, solid state fermentation, submerged fermentation, etc.

E.g. Cellulase (for cellulose hydrolysis), -galactosidase (for lactose hydrolysis, which converts lactose to glucose and galactose) [47],

Cellulase

Research on cellulase has progressed very rapidly in the past few decades', emphasis being in enzymatic hydrolysis of cellulose to glucose [48]. Cellulose constitutes the highest percentage of municipal and wastes. It represents a major source of renewable energy and raw materials. Therefore, the utilization of cellulosic wastes to produce energy is potentially of great importance [49].

Cellulase refers to a class of enzymes produced primarily by fungi, bacteria, and protozoans that catalyze cellulose hydrolysis (cellulolysis). Five types of cellulases based on the kind of reaction catalyzed are as follows

1. ***Endocellulase***: breaks internal bonds to disrupt the crystalline structure of cellulose and expose individual cellulose polysaccharide chains
2. ***Exocellulase***: cleaves two to four units from the ends of the exposed chains produced by endocellulase, resulting in the tetrasaccharides or disaccharides, such as cellobiose. There are two main types of exocellulases [or cellobiohydrolases (CBH)] - CBHI works processively from the reducing end, and CBHII works processively from the nonreducing end of cellulose.
3. ***Cellobiase or beta-glucosidase***: hydrolyses the exocellulase product into individual monosaccharides.

4. ***Oxidative cellulases***: depolymerize cellulose by radical reactions, as for instance cellobiose dehydrogenase (acceptor).
5. ***Cellulose phosphorylases***: depolymerize cellulose using phosphates instead of water [50].

Method of Production

T. reesei is a saprobic fungus capable of efficiently degrading plant cell wall polysaccharides such as cellulose and hemicelluloses. It was first isolated in Solomon Islands in 1944 by the US Army. The original *T. reesei* isolate, *T. reesei* QM6a soon became an area of huge interest, due to its high cellulolytic potential. The construction of different mutant strains with several-fold increase in the amount of secreted cellulolytic enzymes compared to the wildtype strain has been achieved by both academic and industrial research programs.

One of the most hypercellulolytic strains, RUT-C30, originated from *T. reesei* QM6a by three rounds of mutagenesis including a UVlight treatment followed by N-nitrosoguanidine (NTG) and another round of UV-light treatments. Recently, massive parallel sequencing of two mutant strains including RUT-C30 and its ancestor NG-14 (originated from *T. reesei* QM6a) has been reported. Expression of the cellulolytic enzymes in *T. reesei* is induced by cellulose and also by some disaccharides such as D-lactose, cellobiose and sophorose [51]. Cellulase from *T. reesei* can be produced in following steps [52] (Figure 4).

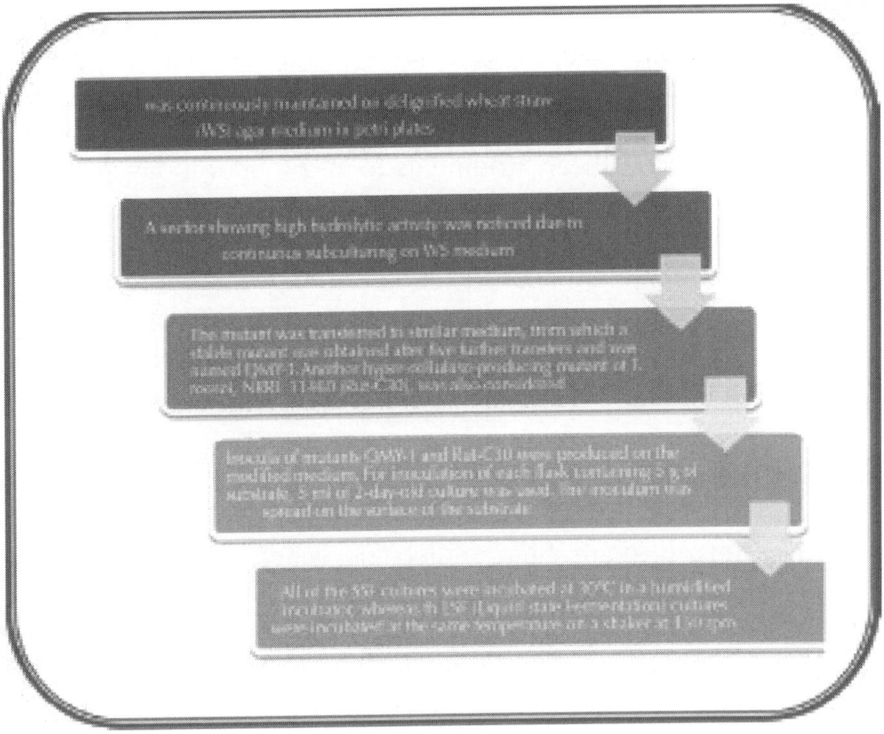

Figure 4: Flow chart for cellulase production by *Trichoderma reesei.*

Cellulases are among the industrially important hydrolytic enzymes and are of great importance in present day biotechnology. Cellulases are widely used in the food, feed, textile and pulp industries [53].

Production of Polysaccharides

During the past several decades, much interest has been generated in the subject of polysaccharides produced by microorganisms, due to their various biological and pharmacological activities, including immuno-stimulating and anti-tumour activities [54]. Polysaccharides are polymeric carbohydrate structures, formed of repeating units (either mono- or di saccharides) joined together by glycosidic bonds. These structures are often linear, but may contain various degrees of branching.

Bacterial polysaccharides represent a varied range of macromolecules that comprise peptidoglycan, bacterial cellulose

[55], lipopolysaccharides, capsules and exopolysaccharides; compounds whose functions range from structural cell-wall components (e.g., peptidoglycan), and chief virulence factors (e.g., Poly-Nacetylglucosamine in *S. aureus*), in order to permit survival of the bacterium in harsh environments (e.g., *Pseudomonas aeruginosa* in the human lung). Polysaccharide biosynthesis is a firmly regulated, energyintensive process and understanding the subtle interplay between the regulation and energy conservation is a huge area of research. The potential benefits are huge and should enable the development of novel antibacterial strategies (e.g., new antibiotics and vaccines) and the commercial exploitation to develop novel applications [56].

Exopolysaccharides

Exopolysaccharides are high molecular weight polymers that are composed of sugar residues and are secreted by a microorganism into the surrounding environment [57]. Microorganisms synthesize a wide range of multifunctional polysaccharides including intracellular polysaccharides, structural polysaccharides and extracellular polysaccharides or exopolysaccharides (EPS). Exopolysaccharides generally consist of monosaccharides and some non-carbohydrate substituents (such as acetate, pyruvate, succinate, and phosphate) [58].

Many dairy lactic acid bacteria which are used for the manufacture of fermented milk products have the ability to produce exopolysaccharides (EPS), which are excreted into the milk during fermentation. They prevent syneresis, and ensure proper texture and body of the end product. Further, it has been reported that EPS isolated from lactic acid bacterial cultures have antitumor activity [59,60]. These lactic acid bacteria also play an important role in the biodegradation of azo dyes [61]. In this article, the production of exopolysaccharides from *Lactobacillus rhamnosus* is discussed.

Method of Production

The exopolysaccharide (EPS)-producing cultures such as *Lactobacillus rhamnosus* RW-9595M present a challenge for the culture producers because the high viscosity of the fermented growth medium make it difficult to recover the cells by centrifugation or filtration [62]. Study has been carried out in order to make it feasible for separation of cells from the culture. Fed-batch culture is observed to be fruitful in this case. Following steps are involved in the production of exopolysaccharides from *Lactobacillus rhamnosus*

Exopolysaccharide (EPS)-producing lactic acid bacteria (LAB) are of importance to the food industry since they can improve the texture of yoghurts, lower the risk of syneresis and improve yields in cheeses. In addition to their technological benefits, LAB EPS have been recognized to have antitumor, antiulcer and blood cholesterol lowering activities, as well as the capability to enhance the immune system. Therefore EPS producing strains are of commercial value for both their technological and putative probiotic properties [62] (Figure 5).

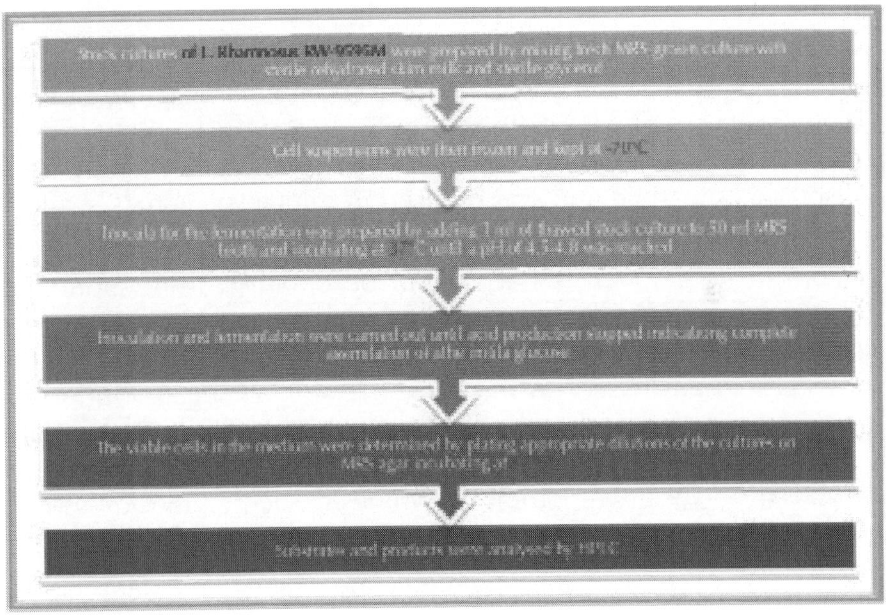

Figure 5: Flow chart of exopolysaccharide synthesis.

Organic Acid Production

An organic acid is an organic compound with acidic properties. The most common organic acids are the carboxylic acids, whose acidity is associated with their carboxyl group –COOH. Simple organic acids such as formic or acetic acids are used for oil and gas well stimulation treatments. These organic acids are much less reactive with metals than strong mineral acids like hydrochloric acid (HCl). Therefore, organic acids are used at high temperatures or when long contact times between acid and pipe are needed. The conjugate bases of organic acids such as citrate and lactate are frequently used in biologically-compatible buffer solutions. Biological systems generate many more complex organic acids such as L-lactic, citric, and D-glucuronic acids that contain hydroxyl or carboxyl groups [63].

In this article, recent development in the production of L-Lactic acid is discussed by New *Lactobacillus Rhamnosus B 103*

Method of Production

Interest in the production of L-lactic acid has increased recently due to its capability to serve as raw material for the manufacture of green solvents, such as ethyl lactate and poly- L-lactic acid (PLLA). These green solvents are biodegradable and environmentally friendly. PLLA is also biocompatible and can be used in implants, such as scaffolds in humans.

Pure lactic acid in its L-form can be obtained from submerged cultures of *Lactobacillus rhamnosus* with a simple and low-cost medium. The biotechnological procedures used for production of lactic acid are based on the bioconversion of sugar solutions by bacteria. Both chemical and biotechnological methods are available for manufacture of lactic acid, but biotechnological production offers several advantages over chemical synthesis, such as the low cost of substrates, milder temperature and less energy consumption.

Lactic acid produced through biotechnological fermentation is also preferred for applications in polymer industries due to the prospects of environmental friendliness and the use of renewable sources rather than petrochemicals [64]. Steps involved in the production of lactic acid from *Lactobacillus Rhamnosus B 103* are shown in Figure 6. Sugarcane plays a key role in agriculture due to its economic and social importance. Because of its rich composition in fermentable sugars, nitrogenous substances (proteins and amino acids) and vitamins as well as its low cost and high availability, sugarcane juice can be used as a substrate for improving lactic acid production and reducing costs [64].

Recombinant Protein Production

Recombinant protein production is the expression of proteins that have been produced by recombinant DNA technology. This process enables these substances to be made in large quantities. Such mass production is done both for laboratory study and for industrial production [65].

High-level production of recombinant proteins as a prerequisite for subsequent purification has become a standard technique. Important applications of recombinant proteins are:

- »≫ Immunization,
- »≫ Biochemical studies,
- »≫ Three-dimensional analysis of the protein, and
- »≫ Biotechnological and therapeutic use.

Production of recombinant proteins involves cloning of an appropriate gene into an expression vector under the control of an inducible promoter. But efficient expression of the recombinant gene depends on a variety of factors such as,

- »≫ Optimal expression signals (both at the level of transcription and translation),
- »≫ Correct protein folding and
- »≫ Cell growth characteristics [66].

One of such recombinant protein on which research was carried out was Human Interferon α.

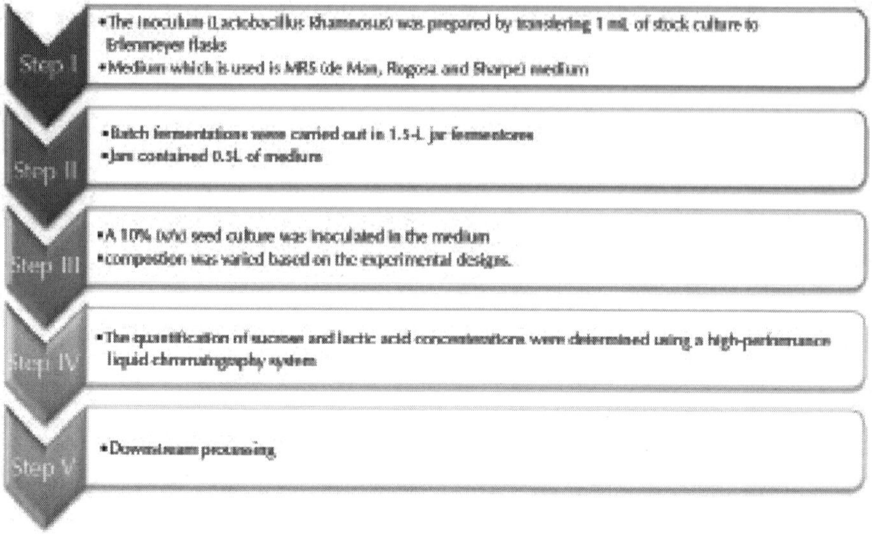

Figure 6: Flowchart for the production of L- lactic acid using *Lactobacillus Rhamnosus*.

Human Interferon α

Interferons (IFNs) are natural cell-signalling proteins produced by the cells of the immune system of most vertebrates in response to challenges such as viruses, parasites and tumor cells. Interferon therapy is used (in combination with chemotherapy and radiation) for the treatment for many cancers, AIDS related Kaposi's sarcoma, and chronic hepatitis B and C.

The expression of IFN- cDNA was achieved directly in *E.coli* soon after it was first cloned. Since, *E. coli* could be grown to high cell densities, and strains used for recombinant protein production are generally regarded as safe, it is a preferred expression host for largescale fermentations [67].

Method of Production

Currently, *Escherichia coli* is one of the main scientific subjects not only for basic research but also for applied biotechnology [68]. Process of production of recombinant proteins first involves the construction of gene which codes for the protein, amplification of that gene, transformation into suitable host (*E.Coli*), growing the colonies on the medium, and then purification of protein. It involves the following steps [67].

- IFNa2b gene was amplified in PCR using the forward primer 5′CCG CCG GGA TCC GAT GAT GAT GAT AAA TGT GAC CTA CCA CAA ACC CAC 3′ that introduces a BamH1 site at the 5′ end of the gene and a reverse primer 5′ CCG CCG GAA TTC AAG CTT TCA TTA CTC TTT AGA TCT TAA 3′containing the EcoR1 site at the 3′ end of the gene
- The ligation mix was then introduced into DH5a competent cells
- Recombinant clones were screened by colony using IFN gene specific primers.
- The PCR positive clones were later selected by restriction digestion and the selected clone was designated as pET21a-IFN
- SAK gene was PCR amplified from the synthetic gene using 5′ CCG CCG GAA TTC CAT ATG TCA TTC GAC AAA GGA 3′ and 5′ CCG CCG GAA TTC TTA TTT ATC ATC ATC ATC GGA TCC TTT CTT TTC TAT AAC AAC 3′ as the forward and reverse primer respectively.
- (SAK will have enterokinase (DDDDK) recognition site.)
- The ligation mix was used for transformation of DH5a cells
- Colonies were screened with colony PCR for SAK gene using gene specific primers.
- The clones were further confirmed by restriction analysis NdeI/EcoRI digestion.
- The construct was designated as pET21a-SAK-IFN
- BL21A1 cells and BL21(DE3) codon plus cells were independently transformed using pET21a- IFN and pET21a- SAKIFN plasmids.
- The cultures were then inoculated into Luria Bertani (LB) with ampicillin at 100 μg/ml and incubated at 37oC till A600nm was nearly 0.5 to 0.6.
- The cells were induced with 13 mM arabinose and 1% lactose in case of BL21A1 cells while 1 mM IPTG was used in case of BL21(DE3) codon plus cells.

- Cells were pelleted, lysed with glass beads and the soluble and insoluble cell fractions were separated
- Suitable aliquots of both the soluble and insoluble fraction were checked for expression on 12% SDS-PAGE gel followed by silver stain.
- Purification of the protein [67]

IFN fusion proteins have been created to render them soluble in *E. coli*. Some of the well-studied interferon fusions include GST-IFN [69], HSA-IFN [70] and the latest report on the fusion of IFN with the antimicrobial peptide as IFN-CM4 fusion [71].

Nowadays there is a rising interest in the development of more efficient and less time-consuming methods to assess the presence of microbes, as well as their viability for bioprocess control and improvement. Rapid detection of microorganisms in samples is one of the importan questions to obtain real-time data for the development of more precise quality control programs, but also for monitoring microbial population during fermentation stages, in order to achieve a better control over processes [73]. Identifications of microbial organisms are now usually done by comparing their SSU rRNA gene sequences to those of known organisms. The standard application is to study the composition of the microbial community within a given environmental or clinical sample [74]. Metabolic profiling is one method which can be used for cataloguing all (or most of) the biochemicals associated with an organism's metabolism [75].

CONCLUSIONS

Fermentation and production of industrially important products involves crucial steps to be followed aseptically. Fermentation is more than 3000 years old and since then, many developments have been made and many implications have been drawn in development of novel techniques for more enhanced improved production of industrially important products. Some of those techniques have been discussed in this review with regard to the production of lactic acid, cellulase, exopolysaccharide, and human hormones. These techniques involved improved strategies like using genetically engineered microbes, and new methodologies which resulted in high yield and cost effective production of these products.

REFERENCES

1. Stanbury PF, Hall SJ, Whitaker A (1998) Principles of fermentation technology, Pergamon Press, Oxford, 1995.
2. Jung EY, Lee HS, Seo HC, Suh HJ (2010) Safety Evaluation of Yeast Hydrolysate (Notress). J Microbial Biochem Technol 2: 58-63.
3. Iyovo GD, Du G, Chen J (2010) Sustainable Bioenergy Bioprocessing: Biomethane Production, Digestate as Biofertilizer and as Supplemental Feed in Algae Cultivation to Promote Algae Biofuel Commercialization. J Microbial Biochem Technol 2: 100-106.
4. http://www.bio-link.org/GMP/Section_1_6_0.html
5. Lee T, Amore TD (2011) Membrane Separation Theoretical and Applicable Considerations for Optimum Industrial Bioprocessing. J Bioprocess Biotechniq 1: 101e
6. http://en.wikipedia.org/wiki/Downstream_processing
7. Miyauchi M, Miao J, Simmons TJ, Dordick JS, Linhardt RJ (2011) Flexible Electrospun Cellulose Fibers as an Affinity Packing Material for the Separation of Bovine Serum Albumin. J Chromatograph Separat Techniq 2: 110.
8. Prasadarao G (2011) TOROCELL- disposable bioreactor system and its applications. J Microbial Biochem Technol. World congress on Biotechnology, India
9. Iyovo GD, Du G, Chen J (2010) Sustainable Bioenergy Bioprocessing: Biomethane Production, Digestate as Biofertilizer and as Supplemental Feed in Algae Cultivation to Promote Algae Biofuel Commercialization. J Microbial Biochem Technol 2: 100-106.
10. Shukla P, Vishwakarma P (2011) Biochemical and Microbial Examination of Sulphi and Cheend: Two Alcoholic Beverages from Central India. J Nutr Food Sci 1: 105.
11. Jemma K & Emma O'Leary , Primary Versus Secondary Metabolites of Microorganisms
12. Benerji DSN, Ayyanna C, Rajini K, Rao BS, Banerjee DRN, et al. (2010) Studies on Physico-Chemical and Nutritional Parameters for the Production of Ethanol from Mahua Flower (Madhuca indica) Using Saccharomyces Cerevisiae – 3090 Through

Submerged Fermentation (smf). J Microbial Biochem Technol 2: 46-50.

13. Dajanta K, Apichartsrangkoon A, Chukeatirote E (2011) Antioxidant properties and total phenolics of thua nao (a Thai fermented soybean) as affected by Bacillus fermentation. J Microbial Biochem Technol 3: 56-59.
14. Gesheva V (2009) Production of Fibrinolytic Enzyme by Streptomyces Rimosus at Conditions of Nitrogen Limitation. J Microbial Biochem Technol 1: 57-58.
15. Vuddaraju SP, Nikku MY, Chaduvula AIR, Dasari VRRK, Donthireddy SRR (2010) Application of Statistical Experimental Designs for the Optimization of Medium Constituents for the Production of L-Asparaginase by Serratia Marcescens. J Microbial Biochem Technol 2: 89-94.
16. Karanam SK, Medicherla NR (2010) Application of Doehlert experimental design for the optimization of medium constituents for the production of L-asparaginase from Palm Kernal cake (Elaeis guineensis). J Microbial Biochem Technol 2: 7-12.
17. Wink PL, Bogdawa HM, Renard G, Chies JM, Basso LA, et al. (2010) Comparison between Two Erwinia carotovora L-Asparaginase II Constructions: cloning, Heterologous Expression, Purification, and Kinetic Characterization. J Microbial Biochem Technol 2: 13-19.
18. Underkofler LA, Barton RR, Rennert SS (1957) Production of Microbial Enzymes and Their Applications
19. Kuric L (2011) Molecular Biocoding of Insulin – Amino Acid Ser. J Bioengineer & Biomedical Sci 1: 102
20. Villela AD, Renard G, Palma MS, Chies JM, Dalmora SL (2010) Human Interferon B1ser17: Coding DNA Synthesis, Expression, Purification and Characterization of Bioactive Recombinant Protein. J Microbial Biochem Technol 2: 111-117.
21. Das KMP, Barve S, Banerjee S, Bandyopadhyay S, Padmanabhan S (2009) A Novel Thermostability Conferring Property of Cherry Tag and its Application in Purification of Fusion Proteins. J Microbial Biochem Technol 1: 59-63.
22. Burdock TJ, Brooks MS, Ghaly AE (2010) A Dehydrogenase Activity Test for Monitoring the Growth of Streptomyces Venezuelae in a Nutrient Rich Medium. J Bioprocess Biotechniq 1: 101

23. Porta R, Di Pierro P, Sorrentino A, Mariniello L (2011) Promising Perspectives for Transglutaminase In "Bioplastics" Production. J Biotechnol Biomaterial 1: 102e.
24. Amin GA (2011) Integrated Two-Stage Process for Biodesulfurization of Model Oil by Vertical Rotating Immobilized Cell Reactor with the Bacterium Rhodococcus erythropolis. J Pet Environ Biotechnol 2: 107.
25. Ukuku D, Yuk HG, Zhang H (2010) Behavior of Pulsed Electric Fields Injured Escherichia coli O157:H7 Cells in Apple Juice Amended with Pyruvate and Catalase. J Microbial Biochem Technol 2: 134-138.
26. Santos RJ Jr, Batista RA, Rodrigues SA, Filho LX, Lima AS (2009) Antimicrobial Activity of Broth Fermented with Kombucha Colonies. J Microbial Biochem Technol 1: 72-78.
27. Iyovo GD, Du G, Chen J (2010) Poultry Manure Digestate Enhancement of Chlorella Vulgaris Biomass Under Mixotrophic Condition for Biofuel Production. J Microbial Biochem Technol 2: 51-57.
28. Daisy A (2011) The Impact and Treatment of Night Soil in Anaerobic Digester: A Review. J Microbial Biochem Technol 3: 43-50.
29. Jiang-ning AI, Bin Z, Jing-ming JIA (2009) The Effects of NO and AgNO3 on Cell Growth and Salidroside Synthesis in Rhodiola sachalinensis A.Bor. Cell Suspension Culture. J Microbial Biochem Technol 1: 11-14.
30. Soorapaneni S, Apte-Deshpande A, Prasad SK, Kumar J, Raiker VA, et al. (2010) Arabinose Promoter Based Expression of Biologically Active Recombinant Human Growth Hormone in E. coli: Strategies for Over Expression and Simple Purification Methods. J Microbial Biochem Technol 2: 38-45.
31. Jensen EB, Carlsen S (1990) Production of recombinant human growth hormone in Escherichia coli: expression of different precursors and physiological effects of glucose, acetate, and salts. Biotechnol Bioeng 36: 1-11
32. Syamsul KMW, Salina MR, Siti Salhah O, Hanina MN, Mohd Basyaruddin AR (2011) Optimization of Lipase Catalyzed Synthesis of Nonyl Caprylate using Response Surface Methodology (RSM). J Biotechnol Biomaterial 1:106.

33. Polychroniadou E, Kanellaki M, Iconomopoulou M, Koutinas AA, Marchant R, et al. (2003) Grape and apple wines volatile fermentation products and possible relation to spoilage. Bioresource Technology 87: 337-339.
34. Selli S, Canbas A,Varlet V, Kelebek H, Prost C, et al. (2008) Characterization of the most odor-active volatiles orange wine made from a Turkish cv. Kozan (Citrussinensis L. Osbeck). J Agric Food Chem 56: 227–234.
35. Dias DR, Schwan RF, Freire ES, Serôdio RD (2007) Elaboration of a fruit wine from cocoa (Theobroma cacao L.) pulp. International Journal of Food Science & Technology 42: 319–329.
36. Kumar YS, Prakasam RS, Reddy OVS (2009) Optimization of fermentation conditions for mango (Mangifera indica L.) wine production by employing response surface methodology. International Journal of Food Science & Technology 44: 2320–2327.
37. Duarte WF, Dias DR, Pereira GVM, Gervásio IM, Schwan RF (2009) Indigenous and inoculated yeast fermentation of gabiroba (Campomanesia pubescens) pulp for fruit wine production. J Ind Microbiol Biotechnol 36: 557–569.
38. Soufleros EH, Pissa P, Petridis D, Lygerakis M, Mermelas K, et al. (2001) Instrumental analysis of volatile and other compounds of Greek kiwi wine, sensory evaluation and optimisation of its composition. Food Chemistry 75: 487–500.
39. Duarte WF, Dias DR, Oliveira JM, Teixeira JA, Silva JBA, et al. (2010) Characterization of different fruit wines made from cacao, cupuassu, gabiroba, jabuticaba and umbu. LWT Food Science and Technology 43: 1564–1572.
40. Sevda SB, Rodrigues L (2011) Fermentative Behavior of Saccharomyces Strains During Guava (Psidium Guajava L) Must Fermentation and Optimization of Guava Wine Production. J Food Process Technol 2:118.
41. Hosseini SM, Khosravi-Darani K (2010) Response Surface Methodology for Mycoprotein Production by Fusarium Venenatum ATCC 20334. J Bioprocess Biotechniq 1: 102
42. Mueller M, Wilkins MR, Banat IM (2011) Production of Xylitol by the Thermotolerant Kluyveromyces marxianus IMB Strains. J Bioprocess Biotechniq 1: 102e

43. Zhang B, Shahbazi A (2011) Recent Developments in Pretreatment Technologies for Production of Lignocellulosic Biofuels. J Pet Environ Biotechnol 2: 108.
44. Gayen S, Ghosh U (2011) Pectinmethylesterase Production from mixed agro- wastes by Penicillium notatum NCIM. 923 in Solid-State fermentation. J Bioremed Biodegrad 2: 119.
45. Ashok P, Selvakumar P, Carlos RS, Poonam N (1999) Solid State Fermentation for The Production of Industrial Enzymes. Current science 77: 149-162
46. http://www.mapsenzymes.com/Making_of_Enzymes.asp
47. Pakizeh M, Namvar-Mahboub M (2011) Experimental Study of Lactose Hydrolysis and Separation in Continuous Stirred Tank -Ultrafiltration Membrane Reactor. J Chem Eng Process Technol 2: 110.
48. Sravani M, Sirisha Ch, Sri Rami reddy D, Yugandhar NM (2011) Production of cellulase by Cellulomonas uda using banana fruit stalk under Solid substrate fermentation. J Biotechnol Biomater. World congress on Biotechnology, India
49. El-Bondkly AM, Aboshosha AAM, Radwan NH, Dora SA (2010) Successive Construction of β-Glucosidase Hyperproducers of Trichoderma Harzianum Using Microbial Biotechnology Techniques. J Microbial Biochem Technol 2: 70-73.
50. http://en.wikipedia.org/wiki/Cellulase
51. Mehdi D, Robert B, Wensheng Q (2011) Effect of different carbon sources on cellulase production by Hypocrea jecorina (Trichoderma reesei) strains. Int J Biochem Mol Biol 2: 274–286.
52. Chahal DS (1984) Solid-State Fermentation with Trichoderma reesei for Cellulase Production. Appl Environ Microbiol 49: 205-210
53. Gilna VV, Khaleel KM (2011) Cellulase Enzyme Activity of Aspergillus Fumigatus from Mangrove Soil on Lignocellulosic Substrate. Recent Research In Science And Technology 3: 132-134
54. Beshay U, Daba A, Gohar Y (2009) Optimization of submerged culture conditions for exo-polysaccharides production by Streptomyces nasri-UV 135 in bioreactor. J Microbial Biochem Technol 1: 43-46.

55. Hungund BS, Gupta SG (2010) Improved Production of Bacterial Cellulose From Gluconacetobacter persimmonis GH-2. J Microbial Biochem Technol 2: 127-133
56. http://en.wikipedia.org/wiki/Polysaccharide
57. Shukla P, Patel N, Mohan Rao R, Shukla J, Verma S, et al. (2011) Isolation and characterization of polyhydroxyalkanoate and exopolysaccharide producing Bacillus sp. PS1 isolated from sugarcane field in Bhilai, India. J Microbial Biochem Technol 3: 33-35.
58. http://en.wikipedia.org/wiki/Exopolysaccharide
59. Cerning J, Renard CMGC, Thibault JF, Bouillanne C, Landon M ,et al. Carbon Source Requirements for Exopolysaccharide Production by Lactobacillus casei CG11 and Partial Structure Analysis of the Polymer. Appl Envir Microbiol 60: 3914 - 3919.
60. Beshay U, Daba A, Gohar Y (2009) Optimization of submerged culture conditions for exo-polysaccharides production by Streptomyces nasri-UV 135 in bioreactor. J Microbial Biochem Technol 1: 43-46.
61. 61. Elbanna K, Hassan G, Khider M, Mandour R (2010) Safe Biodegradation of Textile Azo Dyes by Newly Isolated Lactic Acid Bacteria and Detection of Plasmids Associated With Degradation. J Bioremed Biodegrad 1:112.
62. Claude P, Champagne, Nancy JG, Christophe L (2007) Fermentation technologies for the production of exopolysaccharidesynthesizing Lactobacillus rhamnosus concentrated cultures. Electronic Journal of Biotechnology 10: 211- 220.
63. http://en.wikipedia.org/wiki/Organic_acid.
64. de Lima CJB, Coelho LF, da Silva GP, Alvarez G, Contiero J (2010) L(+) Lactic Acid Production by New Lactobacillus Rhamnosus B 103. J Microbial Biochem Technol 2: 64 69.
65. http://www.wisegeek.com/what-is-recombinant-protein-production.htm
66. Wolfgang S, Ferreira LC (2004) Production of recombinant proteins in Escherichia coli. Genet Mol Biol 27: 442-453.
67. Salunkhe SS, Prasad B, Sabnis-Prasad K, Apte-Deshpande A, Padmanabhan S (2009) Expression and Purification of SAKfused Human Interferon Alpha in Escherichia coli. J Microbial Biochem Technol 1: 5-10.

68. Eremina NS, Yampolskaya TA, Altman IB, Mashko SV, Stoynova NV (2010) Overexpression of ydbK-encoding Putative Pyruvate Synthase Improves L-valine Production and Aerobic Growth on Ethanol Media by an Escherichia coli Strain Carrying an Oxygen-Resistant Alcohol Dehydrogenase. J Microbial Biochem Technol 2: 77-83.

69. Rabhi-Essafi I, Sadok A, Khalaf N, Fathallah DM (2007) A strategy for highlevel expression of soluble and functional interferon alpha as a GST fusion protein in E. coli. Protein Eng Des Sel 20: 201-209.

70. Zhao HL, Xue C, Wang Y, Li XY, Xiong XH, et al. (2007) Circumventing the heterogeneity and instability of human serum albumin-interferon-a2b fusion protein by altering its orientation. J Biotechnol 131: 245-252.

71. Li NN, Liu P, Chen SJ, Zhou LF, Zhang SQ (2009) Construction and expression of a novel bioactive IFN-alpha2b/CM4 fusion protein in E. coli. Microbiol Res (Epub, ahead of Print).

72. Imandi SB, Karanam SK, Garapati HR (2010) Optimization of Process Parameters for the Production of Lipase in Solid State Fermentation by Yarrowia lipolytica from Niger Seed oil Cake (Guizotia Abyssinica). J Microbial Biochem Technol 2: 28-33.

73. da Silva TL, Feijão D, Reis A (2011) Monitoring Rhodotorula glutinis CCMI 145 Stress Physiological Response during Fed-Batch Fermentations Using Multi- Parameter Flow Cytometry. J Microbial Biochem Technol 3: 6-12.

74. Croce O, Chevenet F, Christen R (2010) A New Web Server for the Rapid Identification of Microorganisms. J Microbial Biochem Technol 2: 84-88.

75. Denery JR, Cooney MJ, Li QX (2011) Diauxic and Antimicrobial Growth Phases of Streptomyces Tenjimariensis: Metabolite Profiling and Gene Expression. J Bioengineer & Biomedical Sci 1:101.

Citations

CHAPTER 1

E. Malusá, L. Sas-Paszt, and J. Ciesielska, Technologies for Beneficial Microorganisms Inocula Used as Biofertilizers, doi:10.1100/2012/491206.

CHAPTER 2

Shengjun Xu, Zhihui Bai, Bo Jin, Runlin Xiao, and Guoqiang Zhuang, Bioconversion of wastewater from sweet potato starch production to Paenibacillus polymyxabiofertilizer for tea plants, doi:10.1038/srep04131.

CHAPTER 3

Ashok Kumar Singh, Gauri Singh, Digvijay Gautam, and Manjinder Kaur Bedi, "Optimization of Dairy Sludge for Growth of Rhizobium Cells, doi:10.1155/2013/845264.

CHAPTER 4

Rajiv K. Sinha, Sunita Agarwal, Krunal Chauhan, Vinod Chandran, Brijal Kiranbhai Soni, Vermiculture Technology: Reviving the Dreams of Sir Charles Darwin for Scientific Use of Earthworms in Sustainable Development Programs, doi:10.4236/ti.2010.13019.

CHAPTER 5

Ashok Kumar Singh, Gauri , Rajendra Prasad Bhatt, and Shailja Pant, Optimization and Comparative Study of the Sugar Waste for the Growth of Rhizobium Cells Along with Traditional Laboratory Media, DOI: 10.3923/jm.2011.715.723.

CHAPTER 6

S. K. Yadav, Subhash Babu, M. K. Yadav, Kalyan Singh, G. S. Yadav, and Suresh Pal, A Review of Organic Farming for Sustainable Agriculture in Northern India, http://dx.doi.org/10.1155/2013/718145.

CHAPTER 7

Seema B Sharma, Riyaz Z Sayyed, Mrugesh H Trivedi, and Thivakaran A Gobi Phosphate Solubilizing Microbes: Sustainable Approach For Managing Phosphorus Deficiency In Agricultural Soils, doi: 10.1186/2193-1801-2-587.

CHAPTER 8

Gunindra Nath Chattopadhyay, Use of vermicomposting biotechnology for recycling organic wastes in agriculture, doi: 10.1186/2251-7715-1-8.

CHAPTER 9

Pavan kumar Pindi and SDV Satyanarayana, Liquid Microbial Consortium- A Potential Tool for Sustainable Soil Health, http://dx.doi.org/10.4172/2155-6202.1000124.

CHAPTER 10

Sameera V (2011) Novel Techniques in the Production of Industrially Imperative Products. J Microbial Biochem Technol R1:003. doi:10.4172/1948- 5948.R1-003.

Index

C

D

E

F